The Human Brain during the First Trimester 21- to 23-mm Crown-Rump Lengths

This fourth of 15 short atlases reimagines the classic 5-volume *Atlas of Human Central Nervous System Development.* This volume presents serial sections from specimens between 21 mm and 23 mm with detailed annotations, together with 3D reconstructions. An introduction summarizes human CNS development by using high-resolution photos of methacrylate-embedded rat embryos at a similar stage of development as the human specimens in this volume. The accompanying Glossary gives definitions for all the terms used in this volume and all the others in the *Atlas.*

Key Features

- Classic anatomical atlas
- Detailed labeling of structures in the developing brain offers updated terminology and the identification of unique developmental features, such as, germinal matrices of specific neuronal populations and migratory streams of young neurons
- Appeals to neuroanatomists, developmental biologists, and clinical practitioners
- A valuable reference work on brain development that will be relevant for decades

T0143424

ATLAS OF
HUMAN CENTRAL NERVOUS SYSTEM DEVELOPMENT
Series

Volume 1: The Human Brain during the First Trimester 3.5- to 4.5-mm Crown-Rump Lengths

Volume 2: The Human Brain during the First Trimester 6.3- to 10.5-mm Crown-Rump Lengths

Volume 3: The Human Brain during the First Trimester 15- to 18-mm Crown-Rump Lengths

Volume 4: The Human Brain during the First Trimester 21- to 23-mm Crown-Rump Lengths

Volume 5: The Human Brain during the First Trimester 31- to 33-mm Crown-Rump Lengths

Volume 6: The Human Brain during the First Trimester 40- to 42-mm Crown-Rump Lengths

Volume 7: The Human Brain during the First Trimester 57- to 60-mm Crown-Rump Lengths

Volume 8: The Human Brain during the Second Trimester 96- to 150-mm Crown-Rump Lengths

Volume 9: The Human Brain during the Second Trimester 160- to 170-mm Crown-Rump Lengths

Volume 10: The Human Brain during the Second Trimester 190- to 210-mm Crown-Rump Lengths

Volume 11: The Human Brain during the Third Trimester 225- to 235-mm Crown-Rump Lengths

Volume 12: The Human Brain during the Third Trimester 260- to 270-mm Crown-Rump Lengths

Volume 13: The Human Brain during the Third Trimester 310- to 350-mm Crown-Rump Lengths

Volume 14: The Spinal Cord during the First Trimester

Volume 15: The Spinal Cord during the Second and Third Trimesters and the Early Postnatal Period

The Human Brain during the First Trimester 21- to 23-mm Crown-Rump Lengths

Atlas of Human Central Nervous System Development, Volume 4

Shirley A. Bayer and Joseph Altman

CRC
CRC Press
Taylor & Francis Group
Boca Raton London New York

CRC Press is an imprint of the
Taylor & Francis Group, an **informa** business

First edition published 2023
by CRC Press
6000 Broken Sound Parkway NW, Suite 300, Boca Raton, FL 33487-2742

and by CRC Press
4 Park Square, Milton Park, Abingdon, Oxon, OX14 4RN

CRC Press is an imprint of Taylor & Francis Group, LLC

LCCN no. 2022008216

ISBN: 978-1-032-18331-2 (hbk)
ISBN: 978-1-032-21930-1 (pbk)
ISBN: 978-1-003-27063-8 (ebk)

DOI: 10.1201/9781003270638

Typeset in Times Roman
by KnowledgeWorks Global Ltd.

Access the Support Material at: https://www.routledge.com/9781032183312

CONTENTS

ACKNOWLEDGMENTS

We thank the late Dr. William DeMyer, pediatric neurologist at Indiana University Medical Center, for access to his personal library on human CNS development. We also thank the staff of the National Museum of Health and Medicine, who were at the Armed Forces Institute of Pathology, Walter Reed Hospital, Washington, D.C. when we collected data in 1995 and 1996: Dr. Adrianne Noe, Director; Archibald J. Fobbs, Curator of the Yakovlev Collection; Elizabeth C. Lockett; and William Discher. We are most grateful to the late Dr. James M. Petras at the Walter Reed Institute of Research, who made his darkroom facilities available so that we could develop all the photomicrographs on location rather than in our laboratory in Indiana. Finally, we thank Chuck Crumly, Neha Bhatt, Kara Roberts, Michele Dimont, and Rebecca Condit for expert help during production of the manuscript.

AUTHORS

Shirley A. Bayer received her PhD from Purdue University in 1974 and spent most of her scientific career working with Joseph Altman. She was a professor of biology at Indiana-Purdue University in Indianapolis for several years, where she taught courses in human anatomy and developmental neurobiology while continuing to do research in brain development. Her lengthy publication record of dozens of peer-reviewed scientific journal articles extends back to the mid 1970s. She has co-authored several books and many articles with her late spouse, Joseph Altman. It was her research (published in *Science* in 1982) that proved that new neurons are added to granule cells in the dentate gyrus during adult life, a unique neuronal population that grows. That paper stimulated interest in the dormant field of adult neurogenesis.

Joseph Altman, now deceased, was born in Hungary and migrated with his family via Germany and Australia to the United States. In New York, he became a graduate student in psychology in the laboratory of Hans-Lukas Teuber, earning a PhD in 1959 from New York University. He was a postdoctoral fellow at Columbia University, and later joined the faculty at the Massachusetts Institute of Technology. In 1968, he accepted a position as a professor of biology at Purdue University. During his career, he collaborated closely with Shirley A. Bayer. From the early 1960s to 2016, he published many articles in peer-reviewed journals, books, monographs, and online free books that emphasized developmental processes in brain anatomy and function. His most important discovery was adult neurogenesis, the creation of new neurons in the adult brain. This discovery was made in the early 1960s while he was based at MIT and was largely ignored in favor of the prevailing dogma that neurogenesis is limited to prenatal development. After Dr. Bayer's paper proved that new neurons are adding to granule cells in the hippocampus, his monumental discovery became more accepted. During the 1990s, new researchers "rediscovered" and confirmed his original finding. Adult neurogenesis has recently been proven to occur in the dentate gyrus, olfactory bulb, and striatum through the measurement of Carbon-14—the levels of which changed during nuclear bomb testing throughout the 20th century—in postmortem human brains. Today, many laboratories around the world are continuing to study the importance of adult neurogenesis in brain function. In 2011, Dr. Altman was awarded the Prince of Asturias Award, an annual prize given in Spain by the Prince of Asturias Foundation to individuals, entities, or organizations from around the world who make notable achievements in the sciences, humanities, and public affairs. In 2012, he received the International Prize for Biology, an annual award from the Japan Society for the Promotion of Science (JSPS) for "outstanding contribution to the advancement of research in fundamental biology." This prize is one of the most prestigious honors a scientist can receive. Dr. Altman died in 2016, and Dr. Bayer continues the work they started over 50 years ago. In his honor, she has set up the Altman Prize, awarded each year to an outstanding young researcher in developmental neuroscience by JSPS.

INTRODUCTION

ORGANIZATION OF THE ATLAS

This is the fourth book in the *Atlas of Human Central Nervous System Development* series, 2nd edition. It deals with human brain development during the middle first trimester. The two specimens in this book have crown-rump (CR) lengths from 21–23 mm with estimated gestation weeks (GW) from 8 to 8.4. To link crown-rump lengths to gestation weeks, we relied on ultrasound data in Loughna et al. (2009). These specimens were analyzed in Volume 4 of the 1st edition (Bayer and Altman, 2006). The annotations emphasize the neurogenetic neuroepithelia (NEPs) in the cerebral cortex along the expanding shorelines of the lateral ventricles, the still active neurogenetic NEPs in the basal telencephalon and diencephalon. In the pons and medulla, the NEPs are waning as the parenchyma expands rapidly. The superarachnoid reticulum forms a large capsule around the developing brain. Interactions with sensory axons from the cranial nerves play important roles in shaping the brain.

The present volume features two normal specimens. One is cut in the sagittal plane (C6202), the other (C966) in the frontal/horizontal plane. Each specimen is presented as a series of grayscale photographs of its Nissl-stained nervous system sections including the surrounding body (**Parts II–III**). The photographs are shown from anterior to posterior (frontal/horizontal specimen) and medial to lateral (sagittal specimen). In the sagittal specimen, the left side of each photo is facing anterior, right side posterior, top side dorsal, and bottom side ventral. The dorsal part of each frontal/horizontal photo is toward the top of the page, the ventral part at the bottom, and the midline is in the vertical center. There are extensive computer-aided 3-dimensional reconstructions of the frontal/horizontal brain in C966 showing the overall appearance of the brain and the large ventricular system within it.

SPECIMENS AND COLLECTIONS

The specimens in this book are from the Carnegie Collection in the National Museum of Health and Medicine that was at the Armed Forces Institute of Pathology (AFIP) in Walter Reed Hospital in Washington, D.C. Since the AFIP closed, the National Museum moved to Silver Springs, MD; this collection is still available for research. The Carnegie Collection (designated by a C prefix) started in the Department of Embryology of the Carnegie Institution of Washington, led by Franklin P. Mall (1862-1917), George L. Streeter (1873-1948), and George W. Corner (1889-1981). Over 40-50 years, the specimens are preserved with various fixatives, embedded in paraffin or paraffin/celloidin media, sliced in various cutting planes, and processed with histological stains. Early analyses of specimens were published in the early 1900s in *Contributions to Embryology, The Carnegie Institute of Washington* (now archived in the Smithsonian Libraries). O'Rahilly and Müller (1987, 1994) have given overviews of some specimens.

PLATE PREPARATION

We photograph serial sections of a given specimen at the same magnification with Kodak technical pan black-and-white negative film (#TP442). Develop the film for 6 to 7 minutes in dilution F of Kodak HC-110 developer, stop bath for 30 seconds, Kodak fixer for 5 minutes, Kodak hypo-clearing agent for 1 minute, running water rinse for 10 minutes, and briefly rinse in Kodak photo-flo before drying. Scan the negatives as color positives at 2700 dots per inch (dpi) with a Nikon Coolscan-1000 35 mm negative film scanner attached to a Macintosh PowerMac G3 computer with a plug-in driver built into Adobe Photoshop. Color positive scans bring out more subtle shades of gray. Finally, convert the scans to 300 dpi using the non-resampling method for image size. Using the features of Adobe photoshop, enhance contrast, correct uneven staining, and slightly darken or lighten areas of uneven exposure.

The photos chosen for annotation in **Parts II-III** are presented in companion plates. The *low-magnification plates* of sagitally sectioned C6202 are designated as **A** through **D** on two sets of facing pages. **Part A** on the left page shows the full-contrast photograph of the brain in the skull without labels. **Part B** on the right page shows low-contrast copies of the same photograph with superimposed outlines of brain parts and labels of major brain ventricles and structures. **Part C** on the second left page shows a full-contrast photo of a slightly larger brain "dissected" from its peripheral structures but preserving cranial sensory structures. **Part D** on the second right page is a low-contrast copy of the photo in C with more detailed labeling in the brain. Several *high-magnification plates* feature enlarged views of the brain to show tissue organization. Photos of C966 are presented in one set of companion plates. This presentation allows users to see the entire section as it would appear in a microscope and then consult the detailed markup in the low-contrast copy on the facing page leaving little doubt about what is being identified. The labels themselves are not abbreviated, so there is no lookup on a list. Different fonts are used to label different classes of structures: the ventricular system is labeled in **CAPITALS**, the neuroepithelium and other germinal zones in **Helvetica bold**, transient structures in ***Times bold italic***, and permanent structures in Times Roman or **Times bold**. Adobe Illustrator is used to superimpose labels and to outline structural details on the low-contrast images. Adobe InDesign is used to place plates into book layout. Finally, high-resolution portable document files (pdf) were uploaded to CRC Press/Taylor & Francis websites.

3-DIMENSIONAL COMPUTER RECONSTRUCTIONS

The entire brain and upper cervical spinal cord of C966 is three-dimensionally reconstructed to show the large superventricles and the surface features of the neuroepithelium in the telencephalon, diencephalon, and rhombencephalon. The brain reconstructions require five steps. First, photograph serial sections through the entire brain; scan the negatives, and convert them to computer files. Second, place the files of sections selected for the reconstruction into one large Adobe Photoshop file that contains a separate photograph in each layer. Alter the visibility and transparency of the layers and align the sections to each other as they were before slicing. Export each layer as a separate file. Third, use Adobe Illustrator to outline the brain surface and the edge of the ventricles, and save the outlines in separate Adobe Illustrator eps (encapsulated postscript) files. Fourth, import the eps files into 3D space (x, y, and z coordinates) using Cinema 4DXL (C4D, Maxon Computer, Inc.), a modeling and animation software package. For each section, points on the outlines have unique x-y coordinates and the same z coordinate. By calculating the distance between sections, the entire array of outlines is stretched in the z axis. The C4D loft tool builds aseparate "skin" of spline mesh polygons for the brain surface and for the ventricular surface. The polygons start from the x-y points on the first outline with the most anterior z coordinate, to the x-y points on the next outline behind it, and finish with the x-y points on the last outline at the most posterior z coordinate. Render the polygon skins either as completely opaque or partly transparent surfaces using the C4D ray-tracing engine. Make the surfaces either invisible or visible using the various options in C4D. The complex structure of the brain requires that the surfaces be built in several different lofts to avoid twisting in the regions of the brain flexures. When these different loft segments are shown together, there are a few unavoidable artifacts, such as surface indentations and changes in the way light reflects from the surface; some of these are labeled in **Figures 22–31**. Fifth, convert the rendered images to Adobe Photoshop files. To make the images easier to understand, use Adobe Illustrator to label the structures and draw thin lines on some of the surfaces. Reconstruction of the parts of the neuroepithelium (**Figures 28–31**) follow the same five steps except that only selected parts of the neuroepithelium in each section are outlined and are rendered as completely opaque blocks of tissue.

NEUROGENESIS IN SPECIMENS (CR 21–23 mm)

Embryonic day (E) 16 rat brains have a similar morphological appearance to human brains with CR lengths from 21–23 mm, and we assume that similar developmental events are happening in the two species (Bayer et al., 1993, 1995; Bayer and Altman, 1995). Based on our timetables of neurogenesis using ^{3}H-thymidine dating methods (Bayer and Altman, 1995, 2012-present; Bayer et al., 1993,1995), **Table 1** lists populations being generated in the spinal cord, medulla, pons, and cerebellum (**Table 1A**), the mesencephalon (**Table 1B**), the hypothalamus and preoptic area (**Table 1C**), the thalamus (**Table 1D**), the pallidum/striatum, amygdala, and septum (**Table 1E**), and the cerebral cortex, hippocampus, and olfactory structures (**Table 1F**).

Many neuronal populations have already been generated and are either migrating or settling in the expanding parenchyma, especially in the medulla, pons, mesencephalic tegmentum and pallidum/basal telencephalon (Bayer and Altman, 2012-present). We use methacrylate-embedded rat embryos on E16 to show the fine details of major parts of the brain (**Figs. 1-19**, Bayer, 2013-present) because the preservation of human specimens does not often show great detail. Each table and figure set will be discussed briefly to summarize development in different regions of the brain. All terms used in the annotations are defined in the comprehensive *Glossary* that accompanies the *Atlas* and readers are strongly encouraged to use that.

Table 1A: Neurogenesis by Region

REGION and NEURAL POPULATION	CROWN RUMP LENGTH 21-23 mm
SPINAL CORD	
Dorsal horn interneurons	●●
MEDULLA/PONS	
Salivatory	●●
Cochlear nuclei	●
Trapezoid (medial)	●
Lateral superior olive	●●
Trigeminal (V) principal sensory	●
Infratrigeminal (V)	●●
Raphe complex	●●
Ventral n. lateral lemniscus	●●
Dorsal tegmental nucleus	●
CEREBELLUM and PRECEREBELLAR NUCLEI	
Pontine reticular nucleus	●●
Pontine nuclei	●

Table 1A. Neural populations in the spinal cord, medulla, pons, and cerebellum that are being generated in rats on Embryonic day (E) 16 (comparable to humans at CR 21–23 mm). *Green dots* indicate the amount of neurogenesis occurring: one dot=<15%; two dots=15-90%. This same dot notation is used for all of the remaining parts (**B-F**) of **Table 1** on the following pages.

SPINAL CORD, MEDULLA, PONS and CEREBELLUM

Dorsal horn interneurons (**Table 1A**) are he only neurons still being generated in the spinal cord. Since the spinal cord is not included in photos of the specimens in this volume, we will not illustrate it in this volume, but will do so in Volume 14 of the *Atlas*.

Throughout the medulla and pons, all motor nuclei associated with the cranial somatic motor nerves have been generated. By now, motor neurons have left the NEPs and are actively differentiating. A sagittal section of the midline spinal cord shows the midline raphe glial structure and a column of motor neurons that extends from the pontine flexure (**Fig. 1B**, possibly facial or trigeminal motor neurons) throughout the medulla (**Fig. 1A**). The exception is that the salivatory nucleus is still being generated in the medulla, which is a preganglionic parasympathetic motor nucleus that controls the salivary glands via the facial and glossopharyngeal nerves. It is important to note that these motor neurons are spindle-shaped in the sagittal plane and are harder to detect in the coronal plane. That indicates that they are migrating and there is no separation between motor groups in the motor column. Thus, there may be intermingling of neuronal populations at this time.

Many sensory neuronal structures have already been generated in the medulla, namely the gracilis, cuneatus, solitary, and vestibular nuclei. The lateral trapezoid, medial superior olive, dorsal nucleus of the lateral lemniscus, parabrachial nucleus, and the infratrigeminal area have been generated in the pons. Afferents from sensory cranial ganglia form prominent fiber tracts in the lateral pons and medulla (**Fig. 2**). The remaining neuronal populations still being generated in the medulla and pons are all associated with either the trigeminal or auditory sensory systems or the precerebellar nuclei. The relationship between the often confusing auditory NEP and precerebellar NEP is-clearly shown in a horizontal section through both germinal zones (**Fig. 3**). The precerebellar NEP is active, now generating the pontine reticular nucleus and the pontine nuclei. Neurons of the inferior olive, lateral reticular, and external cuneate nuclei have already been generated and are migrating in the posterior intramural migratory stream (inferior olive) and the posterior extramural migratory stream (lateral reticular and external cuneate). All of these details are illustrated in **Figures 4-5**.

Deep nuclear neurons and Purkinje neurons have already been generated in the cerebellum and are exiting the neuroepithelium to sojourn in various layers of the cerebellar transitional field (**Fig. 6**). Its neuroepithelium is still active, building up progenitors of the Golgi cells and the external germinal layer that has yet to migrate beneath the pia and start producing the microneurons in the cerebellar cortex (basket, stellate, and granule cells). The contents of the cerebellar transitional field (CTF) are hypothesized to contain the following: CTF 1 contains fibers that are afferent to the cerebellum, possibly from the spinal cord. CTF2 contains sojourning deep nuclear neurons that are spindle-shaped in the sagittal plane and are probably migrating. CTF3 contains fibers also afferent to the cerebellum from the spinal cord or perhaps even the vestibular ganglion. CTF4 contains spindle-shaped migrating and sojourning deep nuclear neurons. CTF5 is an intermittent cell-sparse layer with fibers. CTF6 blends with the basal part of the cerebellar neuroepithelium and contains sojourning Purkinje neurons that will only continue their radial migration to the cerebellar surface when the external germinal layer starts to grow from back to front beneath the pia of the cerebellum.

THE RAPHE GLIAL STRUCTURE IN AN E16 RAT EMBRYO

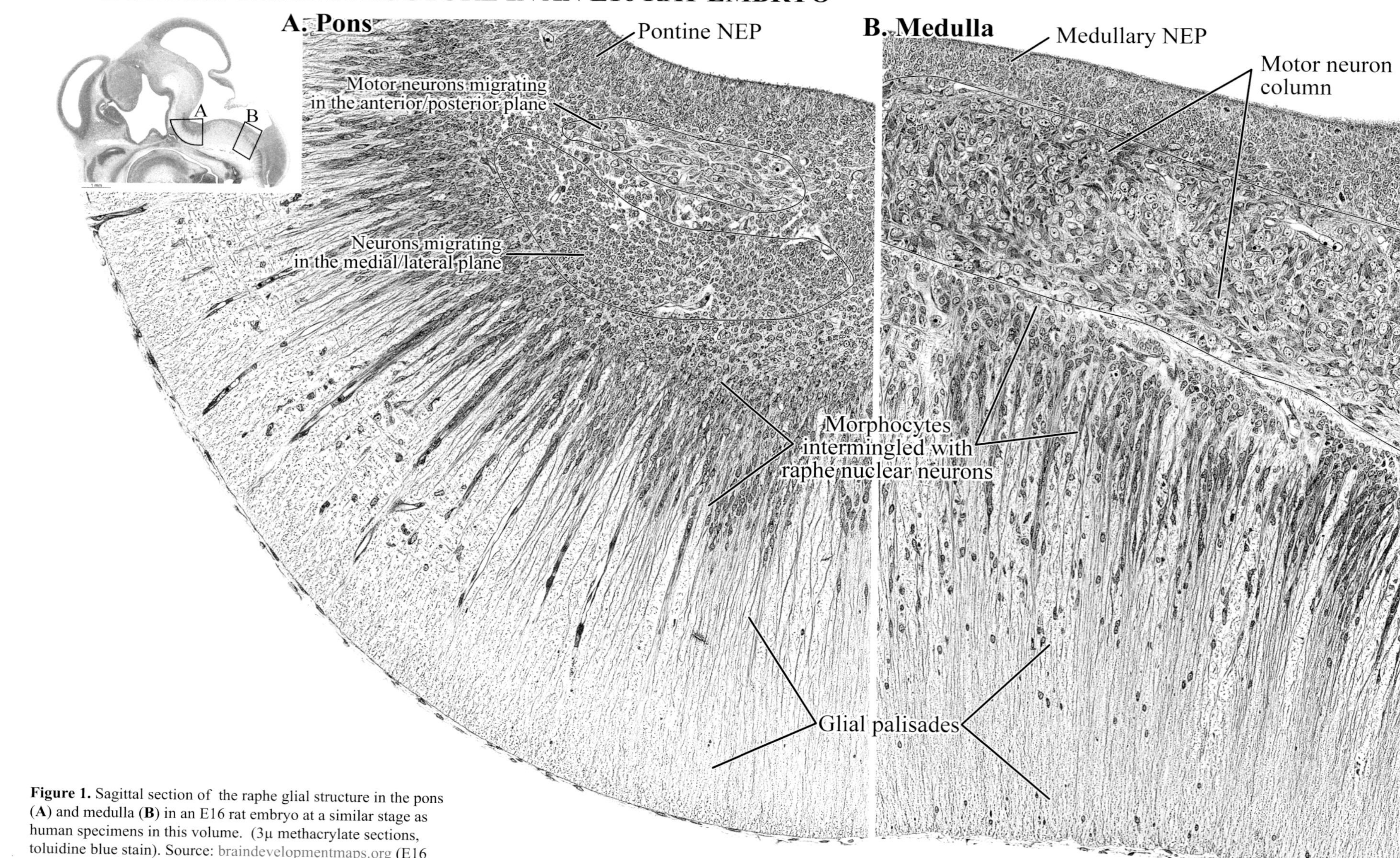
A. Pons
B. Medulla
Pontine NEP
Medullary NEP
Motor neuron column
Motor neurons migrating in the anterior/posterior plane
Neurons migrating in the medial/lateral plane
Morphocytes intermingled with raphe nuclear neurons
Glial palisades
NEP-Neuroepithelium
0.1 mm
A
B

Figure 1. Sagittal section of the raphe glial structure in the pons (**A**) and medulla (**B**) in an E16 rat embryo at a similar stage as human specimens in this volume. (3μ methacrylate sections, toluidine blue stain). Source: braindevelopmentmaps.org (E16 sagittal archive)

THE CEREBELLUM AND SENSORY NERVE ENTRY ZONES IN THE LATERAL PONS AND MEDULLA

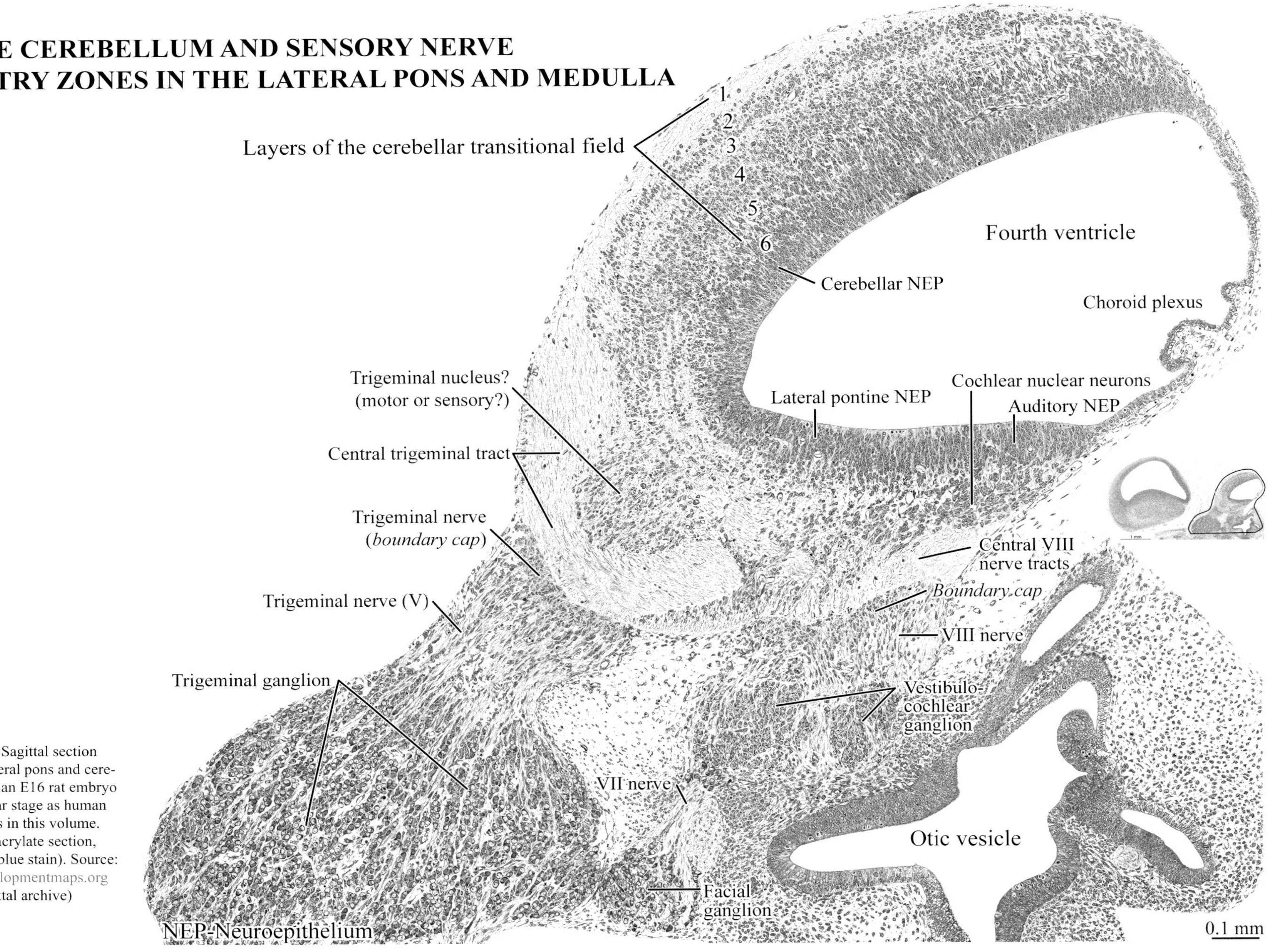

Layers of the cerebellar transitional field
1
2
3
4
5
6
Cerebellar NEP
Fourth ventricle
Choroid plexus
Trigeminal nucleus? (motor or sensory?)
Lateral pontine NEP
Cochlear nuclear neurons
Auditory NEP
Central trigeminal tract
Trigeminal nerve (*boundary cap*)
Central VIII nerve tracts
Boundary cap
Trigeminal nerve (V)
VIII nerve
Trigeminal ganglion
Vestibulo-cochlear ganglion
VII nerve
Otic vesicle
Facial ganglion
NEP-Neuroepithelium
0.1 mm

Figure 2. Sagittal section of the lateral pons and cerebellum in an E16 rat embryo at a similar stage as human specimens in this volume. (3μ methacrylate section, toluidine blue stain). Source: braindevelopmentmaps.org (E16 sagittal archive)

THE AUDITORY AND PRECEREBELLAR NEUROEPITHELIA (NEPs) IN THE HORIZONTAL PLANE

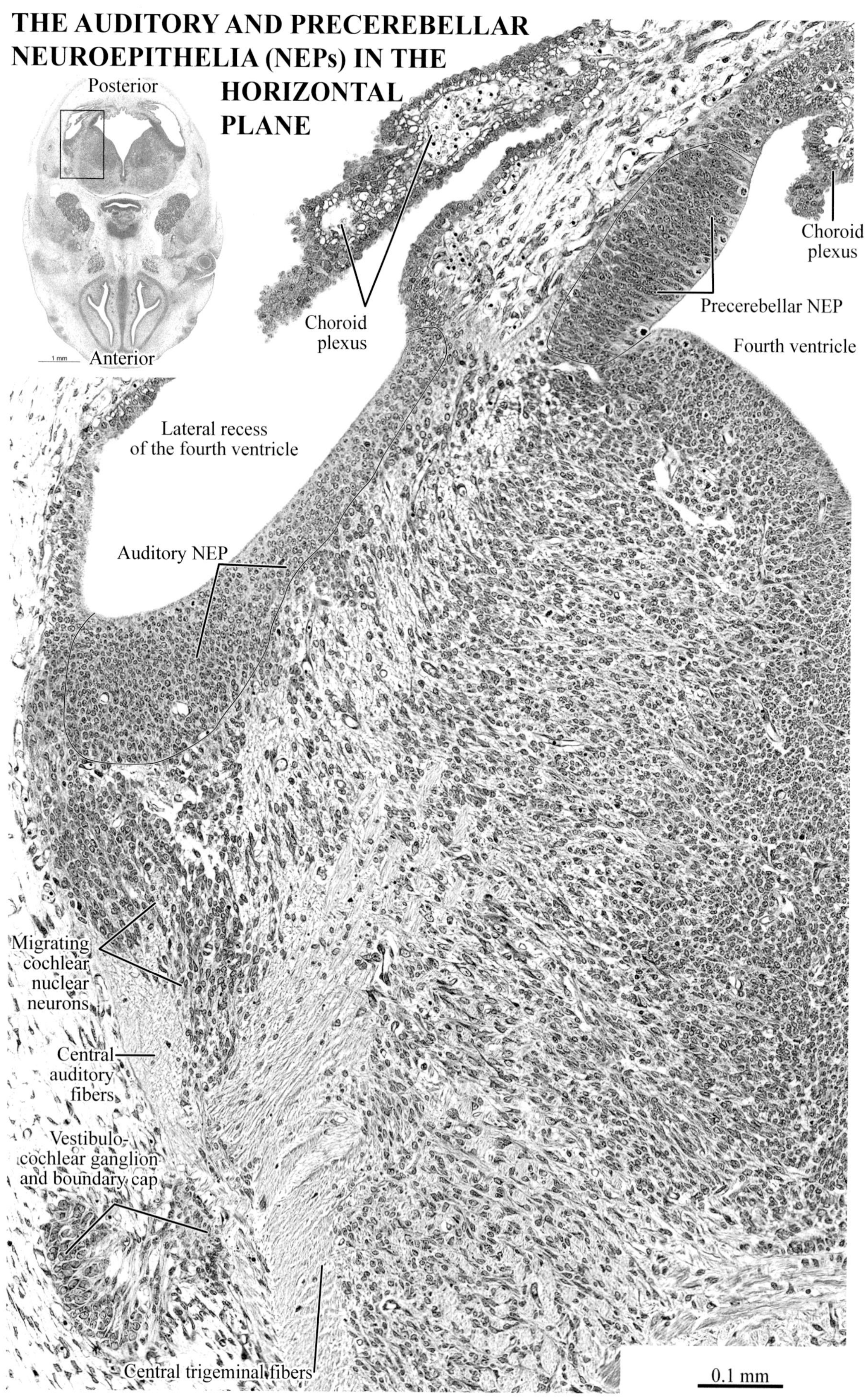
Posterior
Anterior
1 mm
Choroid plexus
Choroid plexus
Precerebellar NEP
Fourth ventricle
Lateral recess of the fourth ventricle
Auditory NEP
Migrating cochlear nuclear neurons
Central auditory fibers
Vestibulo-cochlear ganglion and boundary cap
Central trigeminal fibers
0.1 mm

THE PRECEREBELLAR NEUROEPITHIUM (NEP) IN THE FRONTAL/ HORIZONTAL PLANE

Figure 3. Horizontal section of the medulla showing the auditory and precerebellar neuroepithelia in an E16 rat embryo at a similar stage as human specimens in this volume. (3μ methacrylate section, toluidine blue stain). Source: brain-developmentmaps.org (E16 horizontal archive)

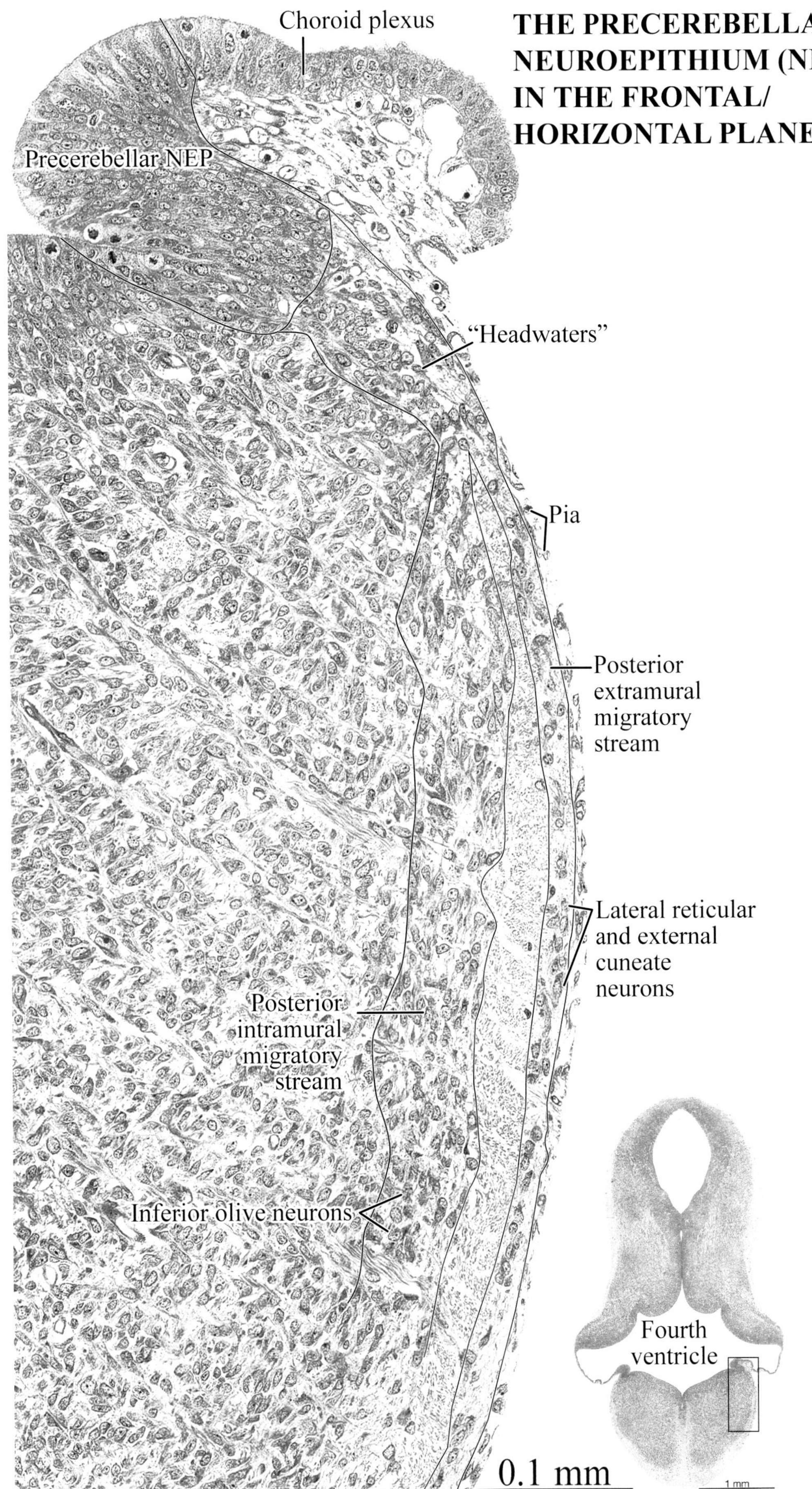
Choroid plexus
Precerebellar NEP
"Headwaters"
Pia
Posterior extramural migratory stream
Lateral reticular and external cuneate neurons
Posterior intramural migratory stream
Inferior olive neurons
Fourth ventricle
0.1 mm
1 mm

Figure 4. Frontal/horizontal section showing detail of the precerebellar neuroepithelium and the "headwaters" of the posterior intramural and extramural migratory streams in an E16 rat embryo at a similar stage as human specimens in this volume. (3μ methacrylate section, toluidine blue stain). Source: brain-developmentmaps.org (E16 coronal archive)

THE PRECEREBELLAR NEUROEPITHIUM IN THE SAGITTAL PLANE

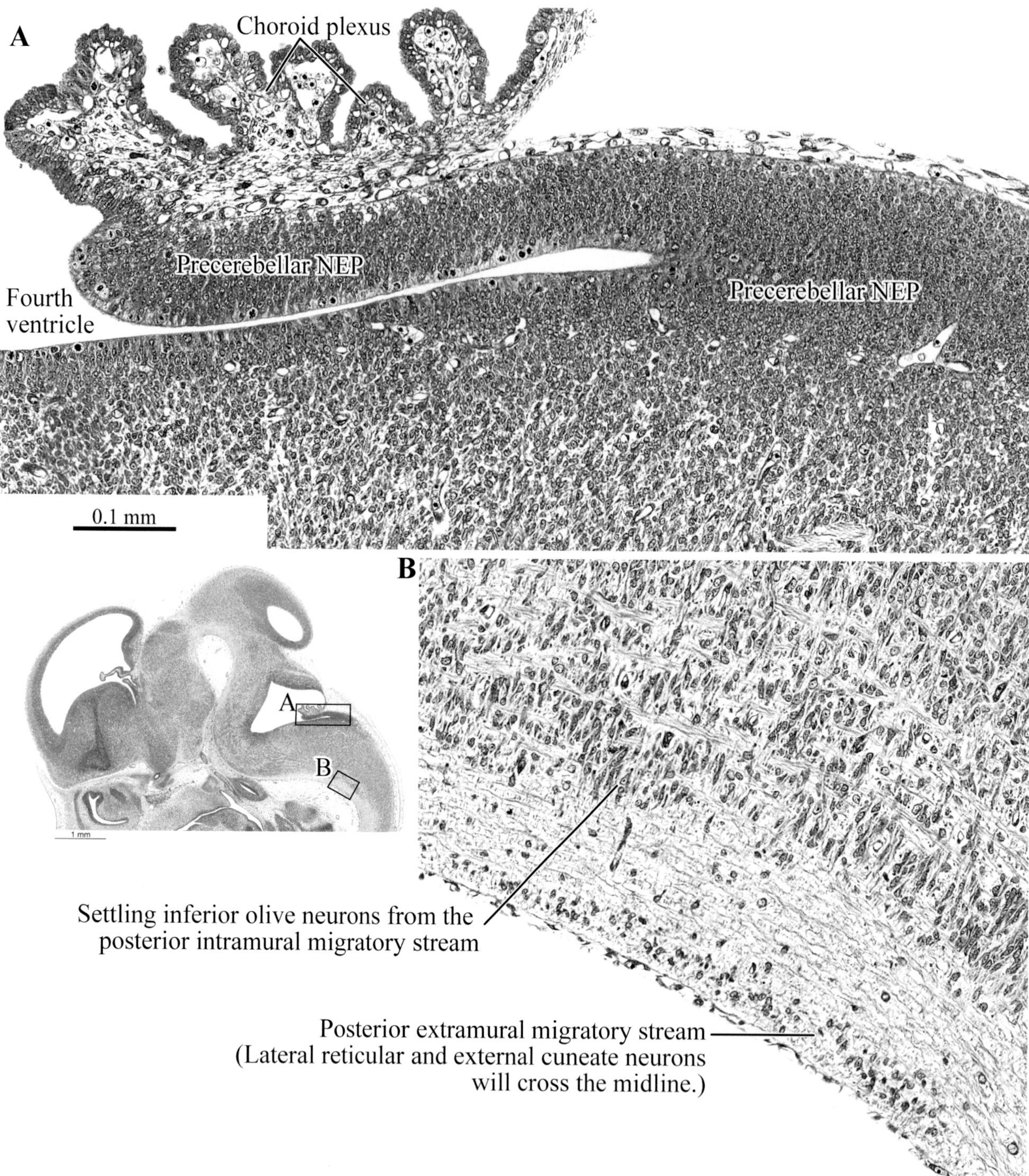
A
Choroid plexus
Precerebellar NEP
Fourth ventricle
Precerebellar NEP
0.1 mm
B
A
B
1 mm
Settling inferior olive neurons from the posterior intramural migratory stream
Posterior extramural migratory stream
(Lateral reticular and external cuneate neurons will cross the midline.)

Figure 5. Sagittal section showing the precerebellar neuroepithelium in an E16 rat embryo at a similar stage as human specimens in this volume. **A**, Detail of the curved elaboration of the precerebellar neuroepithelium (NEP) in the dorsal medulla. **B**, Detail of neurons settling in the inferior olive (from the anterior intramural migratory stream) and neurons that will migrate across the midline on their way to the lateral reticular and external cuneate nuclei (from the extramural migratory stream). (3μ methacrylate section, toluidine blue stain) braindevelopmentmaps.org (E6 sagittal archive)

THE MIDLATERAL CEREBELLUM

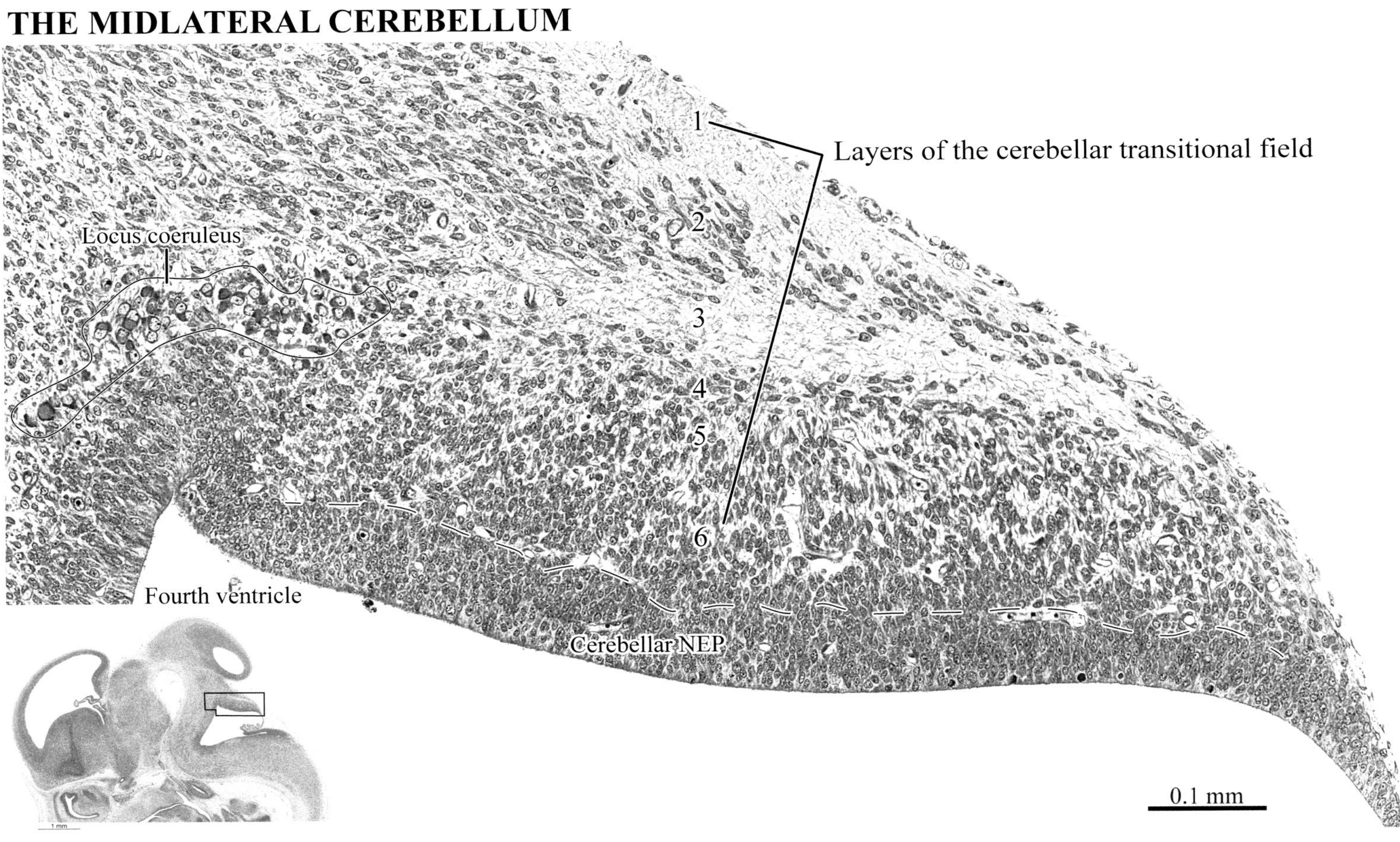
1
2
3
4
5
6
Layers of the cerebellar transitional field
Locus coeruleus
Fourth ventricle
Cerebellar NEP
0.1 mm

Figure 6. Sagittal section showing detail of the cerebellum in an E16 rat embryo at a similar stage as human specimens in this volume. The large neurons at the anterior tip of the cerebellum are in the locus coeruleus, an early-generated (finished in human specimens with crown rump lengths of 15 mm) group of neurons that projects noradrenergic axons to nearly all structures in the brain. (3μ methacrylate section, toluidine blue stain) braindevelopmentmaps.org (E16 sagittal archive)

THE MIDBRAIN TEGMENTUM

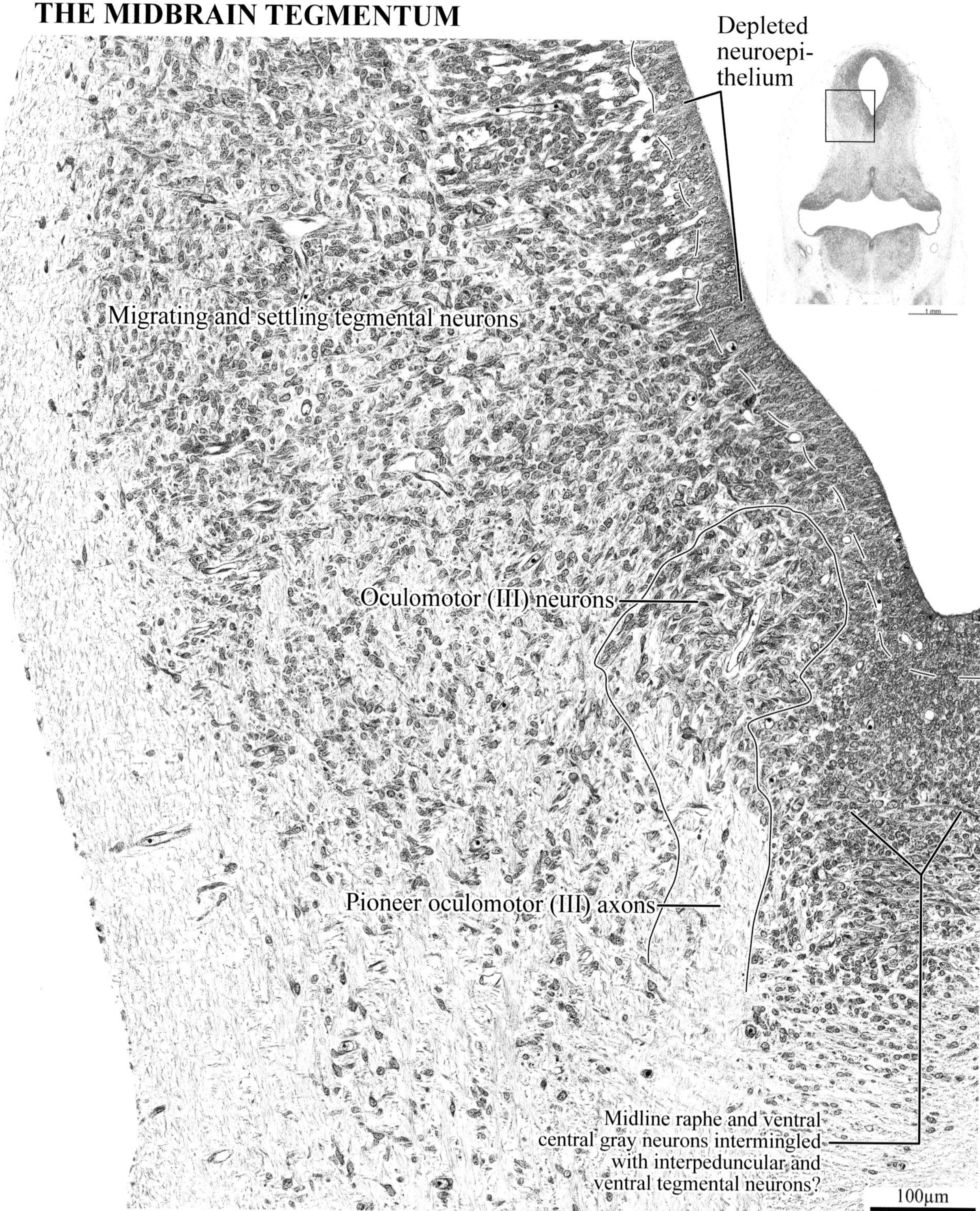
Depleted neuroepithelium
1 mm
Migrating and settling tegmental neurons
Oculomotor (III) neurons
Pioneer oculomotor (III) axons
Midline raphe and ventral central gray neurons intermingled with interpeduncular and ventral tegmental neurons?
100µm

Figure 7. Frontal/horizontal section through the midbrain tegmentum in an E16 rat embryo at a similar stage of development as the human specimens in this volume. Note the depleted neuroepithelium indicating that neurogenesis is nearing completion at this time. (3 µ methacrylate section, toluidine blue stain) Source: braindevelopmentmaps.org (E16 coronal archive)

MIDBRAIN

The midbrain tegmental neuroepithelium is thin because only a few neuronal populations are being generated at this time (**Table 1B**). All motor nuclei in the tegmentum (oculomotor, Edinger-Westphal, Darkschiewitsch, and trochlear) have already been generated and are differentiating. The large motor neurons in the oculomotor nucleus are shown in **Figure 7**. Besides motor structures, neurons in the red nucleus, reticular formation, substantia nigra, lateral ventral tegmental area, ventral central gray, parabi-

Table 1B: Neurogenesis by Region REGION and NEURAL POPULATION	CROWN RUMP LENGTH 21-23 mm
MESENCEPHALIC TEGMENTUM/ISTHMUS	
Ventral tegmental area (medial)	●●
Lateral central gray	●
Dorsal central gray	●●
SUPERIOR COLLICULUS	
stratum album	●
stratum griseum profundum	●●
stratum lemnisci	●●
stratum griseum intermediate	●●
stratum opticum	●●
stratum griseum superficial	●●
stratum zonale	●●
INFERIOR COLLICULUS	
Anterolateral	●●
Posterolateral	●●
Anterior intermediate	●●
Posterior intermediate	●
Anteromedial	●

geminal nucleus, and mesencephalic raphe have also been generated and are migrating and settling in the enlarging parenchyma (**Fig. 7**).

The pretectal germinal zone is still thick but considerably less so than the tectal neuroepithelium (*compare* **Figs. 8A** and **B**). The fibers of the posterior commissure dominate the pretectum and the precursors in its germinal zone are probably producing glia that will surround the fibers there. The tectal neuroepithelium (superior colliculus is shown) is in peak neurogenetic phase (**Fig. 8B**). Neurons are being generated for every layer (**Table 1B**). Cells differentiating and migrating outside the neuroepithelium are mainly destined to settle in the intermediate magnocellular layer (finished at an earlier stage). But many neurons have already been generated during the previous stage in human specimens with crown rump lengths between 16.8 and 17.5 mm (Bayer and Altman 2022, Volume 3, Table 1C). There is no indication of laminar differentiation at this time. The inferior colliculus region of the tectum is just entering its neurogenetic stage (**Table 1B**) and few to no cells are outside the neuroepithelium.

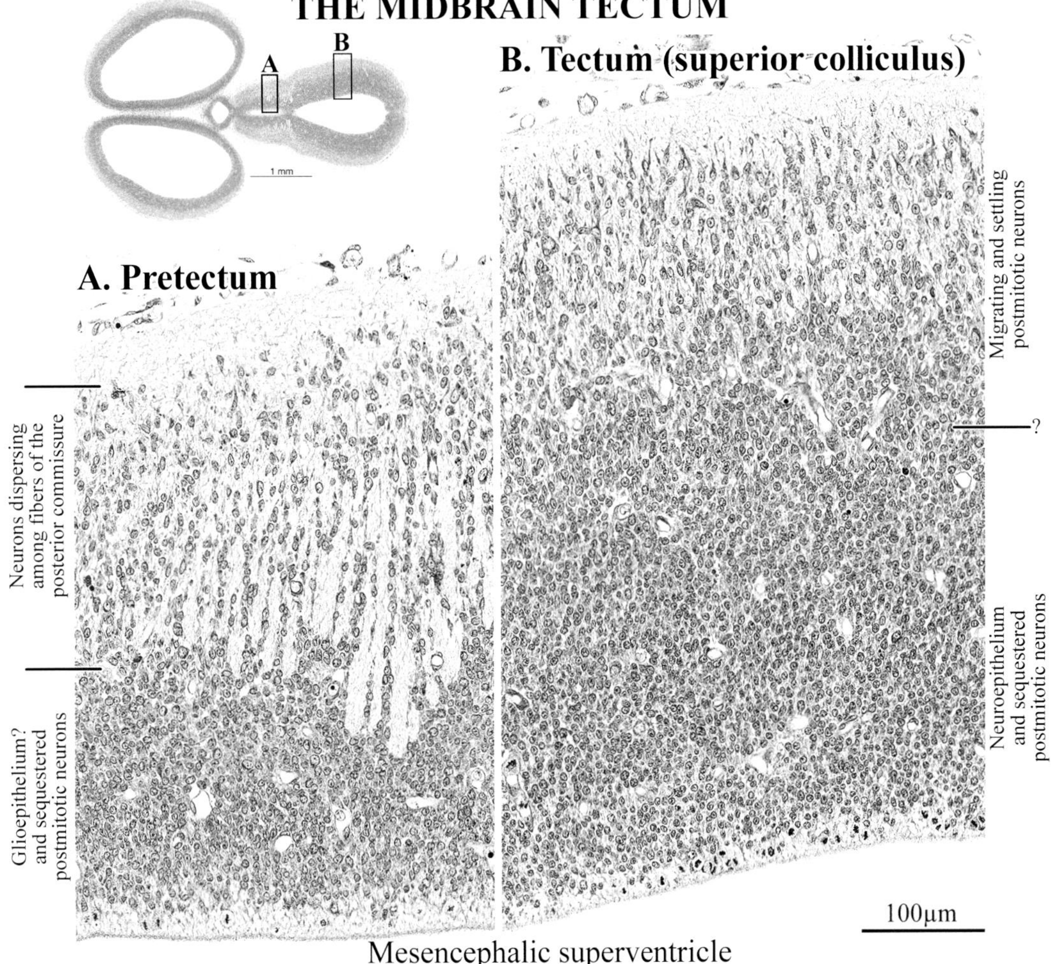
THE MIDBRAIN TECTUM
A
B
1 mm
B. Tectum (superior colliculus)
A. Pretectum
Migrating and settling postmitotic neurons
?
Neurons dispersing among fibers of the posterior commissure
Glioepithelium? and sequestered postmitotic neurons
Neuroepithelium and sequestered postmitotic neurons
100µm
Mesencephalic superventricle

Figure 8. Horizontal section through the pretectum and tectum in an E16 rat embryo at a similar stage of development as the human specimens in this volume. (3 µ methacrylate section, toluidine blue stain) braindevelopmentmaps.org (E16 horizontal archive)

Table 1C: Neurogenesis by Region	
REGION and NEURAL POPULATION	CROWN RUMP LENGTH 21-23 mm
PREOPTIC AREA/ HYPOTHALAMUS	
Medial preoptic area	●
Medial preoptic nucleus	●●
Sexually dimorphic nucleus	●●
Periventricular preoptic nucleus	●●
Median preoptic nucleus	●
Ventromedial nucleus	●
Dorsomedial nucleus	●●
Arcuate nucleus	●●
Suprachiasmatic nucleus	●●
Suprammillary nucleus	●●
Tuberommillary nucleus	●●
Medial mammillary n. (ventral)	●●

Table 1D: Neurogenesis by Region	
REGION and NEURAL POPULATION	CROWN RUMP LENGTH 21-23 mm
THALAMUS	
Anterodorsal	●●
Anteroventral	●●
Anteromedial	●●
Lateral dorsal	●●
Medial Dorsal	●●
Paraventricular	●
Paratenial	●●
Parafascicular	●●
Reuniens	●●
Rhomboid	●●
Ventral complex (VA/VL)	●
Ventroposterior (medial)	●●
Lateral habenula	●
Medial habenula	●●

DIENCEPHALON

Neurogenesis throughout the preoptic area and hypothalamus (**Table 1C**) is robust in medial structures. The lateral preoptic and lateral hypothalamic areas have already been generated, have migrated into the parenchyma, and are differentiating there. Besides that, the supraoptic and paraventricular neurosecretory neurons have been generated along with the premammillary, lateral mammillary and dorsal part of the medial mammillary nucleus. In the thalamus (**Table 1D**), neurogenesis is robust in many nuclear groups, including the anterior complex and medial nuclei. The ventral complex is still being generated in the intermediate lobule and habenular neurons are still originating. The thalamic reticular nucleus has been generated as well as posterior complex neurons in the geniculate bodies. Lateral and medial geniculate neurons are migrating from their germinal source in the superior lobule to meet incoming optic fibers and auditory fibers, respectively (**Fig. 9B**).

We have chosen to illustrate diencephalic development in a series of four photos of a methacrylate-embedded rat embryo on E16 (comparable to human CRs between 21 and 23 mm) in a frontal/horizontal section that slices the diencephalon at the level of the pituitary gland (**Figs. 9-12**) and a frontal/horizontal section of the eye (**Fig. 13**) showing pioneer fibers leaving the retina and heading toward the future optic chiasm.

THE DIENCEPHALON

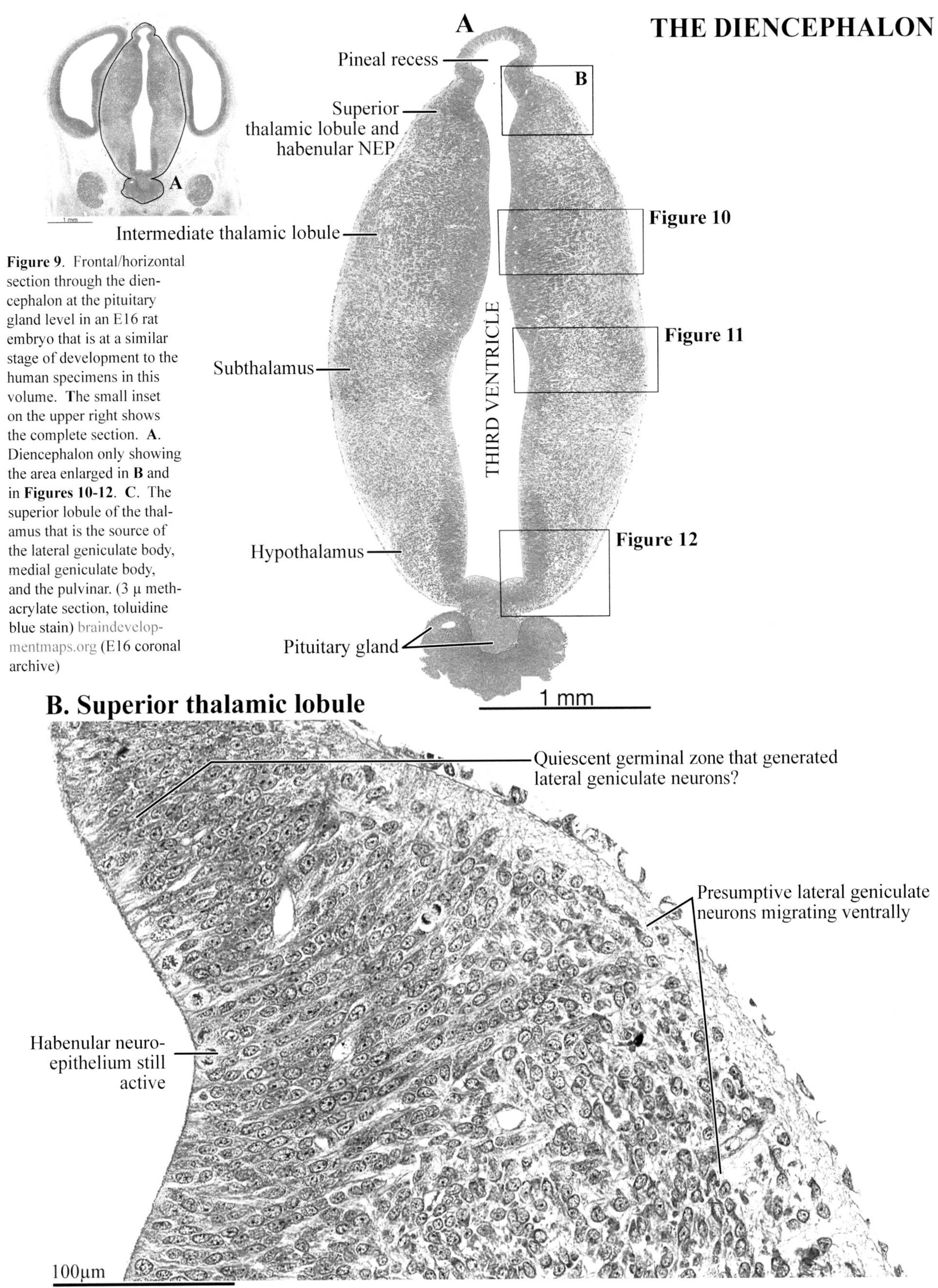
A
A
Pineal recess
B
Superior
thalamic lobule and
habenular NEP
Intermediate thalamic lobule
Figure 10
Figure 11
THIRD VENTRICLE
Subthalamus
Hypothalamus
Figure 12
Pituitary gland
1 mm
1 mm
B. Superior thalamic lobule
Quiescent germinal zone that generated
lateral geniculate neurons?
Presumptive lateral geniculate
neurons migrating ventrally
Habenular neuro-
epithelium still
active
100µm

Figure 9. Frontal/horizontal section through the diencephalon at the pituitary gland level in an E16 rat embryo that is at a similar stage of development to the human specimens in this volume. The small inset on the upper right shows the complete section. **A**. Diencephalon only showing the area enlarged in **B** and in **Figures 10-12**. **C**. The superior lobule of the thalamus that is the source of the lateral geniculate body, medial geniculate body, and the pulvinar. (3 µ methacrylate section, toluidine blue stain) braindevelopmentmaps.org (E16 coronal archive)

THE INTERMEDIATE THALAMIC LOBULE

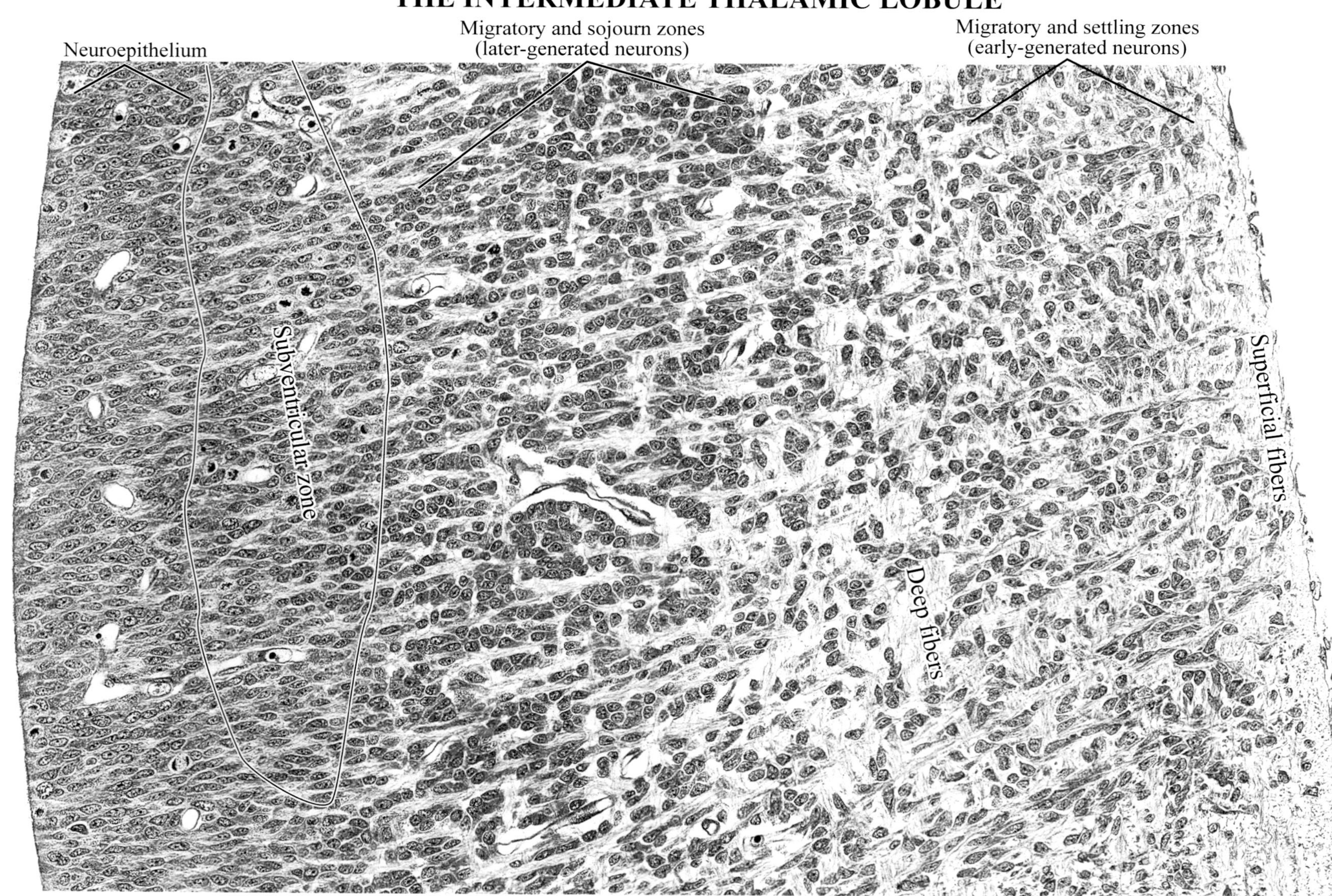
Neuroepithelium
Migratory and sojourn zones
(later-generated neurons)
Migratory and settling zones
(early-generated neurons)
Subventricular zone
Deep fibers
Superficial fibers
100µm

Figure 10. Frontal/horizontal section through the intermediate lobule of the thalamus from the area indicated in **Figure 9**. There are sparse mitotic figures in a subventricular zone. This part of the thalamus is the primordium of the ventral complex (VA/VL and VBA/VBL). The parenchyma shows superficial and deep fibers intermingling with sojourning, migrating and settling neurons. (3 µ methacrylate section, toluidine blue stain) Source: braindevelopmentmaps.org (E16 coronal archive)

THE SUBTHALAMUS

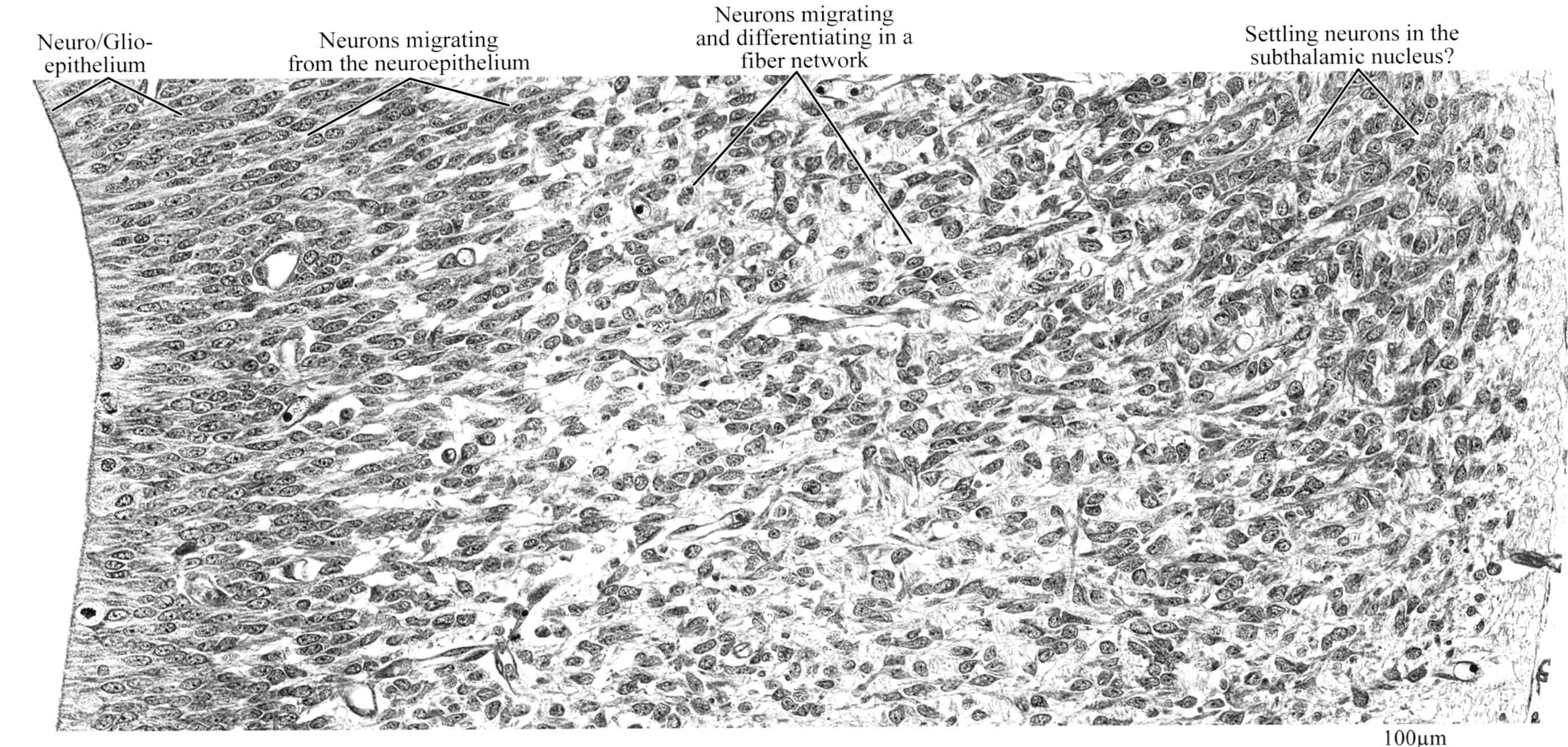
Neuro/Glio-
epithelium
Neurons migrating
from the neuroepithelium
Neurons migrating
and differentiating in a
fiber network
Settling neurons in the
subthalamic nucleus?
100µm

Figure 11. Frontal/horizontal section through the subthalamus from the indicated area of the section in **Figure 9**. Neurogenesis is complete in the subthalamus, but there are mitotically active cells near the ventricular lumen. The cells generated here may be neurons destined to migrate and settle elsewhere, or (more likely) glia to support subthalamic neuronal populations. The neurons settling in the presumptive subthalamic nucleus migrate in from the hypothalamus. (3 µ methacrylate section, toluidine blue stain) Source: braindevelopmentmaps.org (E16 coronal archive)

THE HYPOTHALAMUS

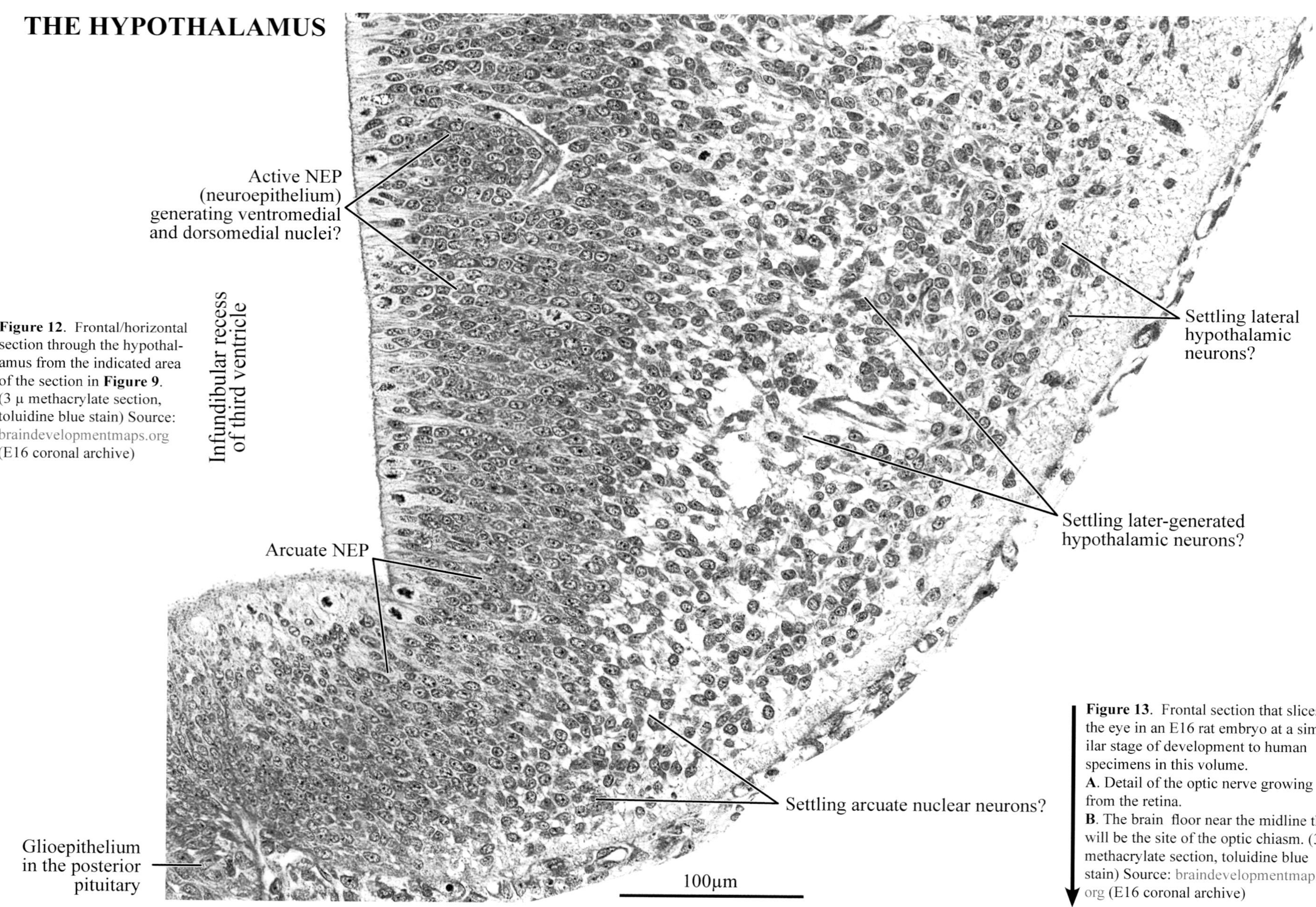
Active NEP
(neuroepithelium)
generating ventromedial
and dorsomedial nuclei?
Infundibular recess
of third ventricle
Settling lateral
hypothalamic
neurons?
Settling later-generated
hypothalamic neurons?
Arcuate NEP
Settling arcuate nuclear neurons?
Glioepithelium
in the posterior
pituitary
100μm

Figure 12. Frontal/horizontal section through the hypothalamus from the indicated area of the section in **Figure 9**. (3 μ methacrylate section, toluidine blue stain) Source: braindevelopmentmaps.org (E16 coronal archive)

Figure 13. Frontal section that slices the eye in an E16 rat embryo at a similar stage of development to human specimens in this volume.
A. Detail of the optic nerve growing from the retina.
B. The brain floor near the midline that will be the site of the optic chiasm. (3 μ methacrylate section, toluidine blue stain) Source: braindevelopmentmaps.org (E16 coronal archive)

THE OPTIC NERVE AND CHIASMATIC REGION

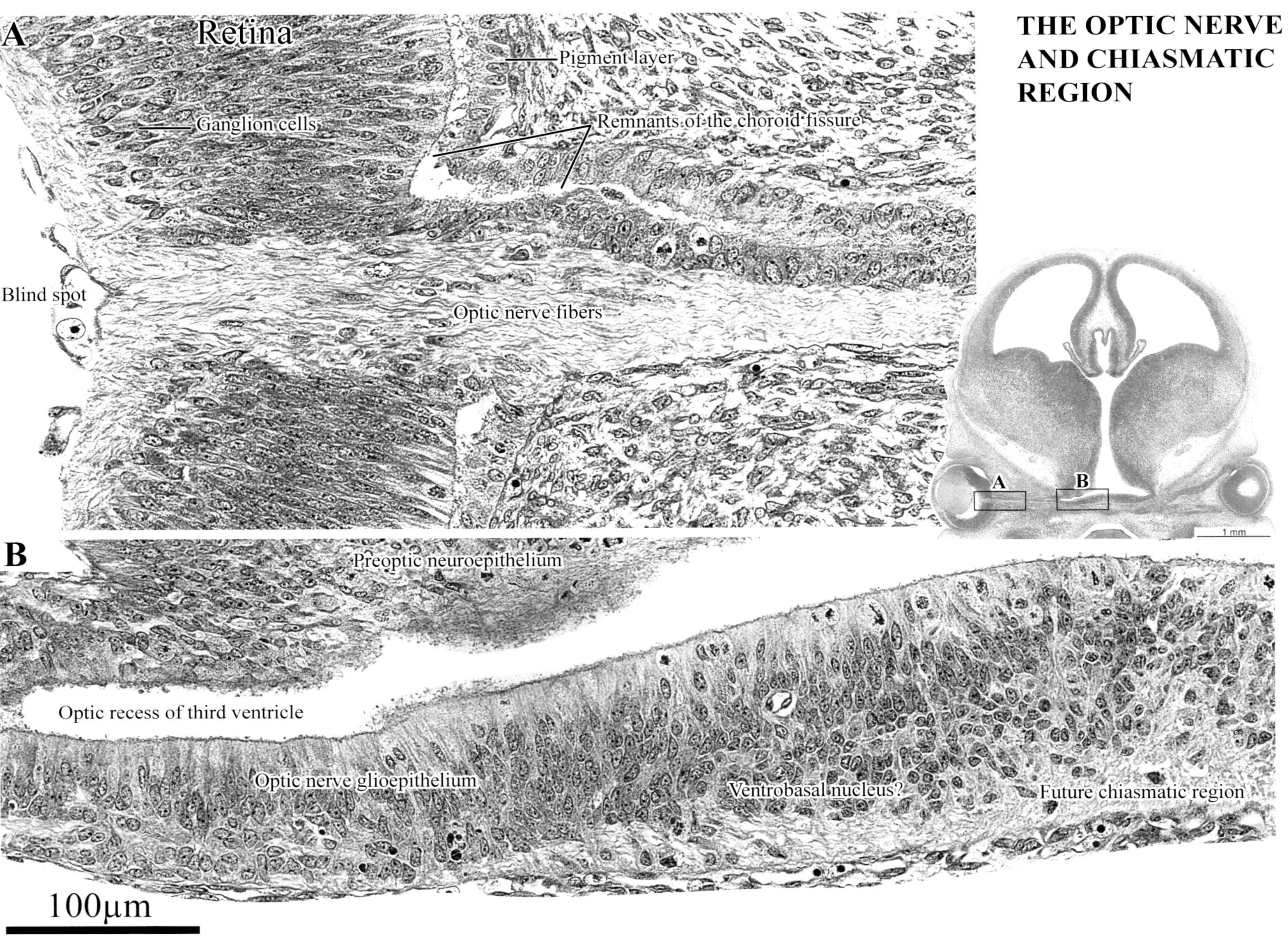
A
Retina
Ganglion cells
Pigment layer
Remnants of the choroid fissure
Blind spot
Optic nerve fibers
A
B
1 mm
B
Preoptic neuroepithelium
Optic recess of third ventricle
Optic nerve glioepithelium
Ventrobasal nucleus?
Future chiasmatic region
100µm

Table 1E: Neurogenesis by Region

REGION and NEURAL POPULATION	CROWN RUMP LENGTH 21-23 mm
PALLIDUM AND STRIATUM	
Globus pallidus (external segment)	●
Substantia innominata	● ●
Basal nucleus of Meynert	● ●
Olfactory tubercle (large neurons)	● ●
Olfactory tubercle (small neurons)	● ●
Caudate and putamen	●
Nucleus accumbens	●
Islands of Calleja	●
AMYGDALA	
Nucleus of the lateral olfactory tract	●
Central nucleus	●
Intercalated masses	● ●
Amygdalo-hippocampal area	●
Medial nucleus	●
Anterior cortical nucleus	●
Posterior cortical nucleus	● ●
Basomedial nucleus	● ●
Basolateral nucleus	●
Lateral nucleus	● ●
Bed n. stria terminalis (anterior)	● ●
Bed n. stria terminalis (preoptic continuation)	● ●
SEPTUM	
Medial nucleus	●
Diagonal band (vertical limb)	● ●
Triangular nucleus	● ●
Lateral nucleus	● ●
Bed nucleus of the anterior commissure	● ●

Table 1F: Neurogenesis by Region

REGION and NEURAL POPULATION	CROWN RUMP LENGTH 21-23 mm
NEOCORTEX and LIMBIC CORTEX	
Cajal-Retzius neurons	●
Layer V	● ●
Layer VI	● ●
Layera IV-II	●
Subplate VII	●
OLFACTORY CORTEX	
Layer II (anterior)	● ●
Layer II (posterior)	● ●
Layers III-IV (anterior)	● ●
Layers III-IV (posterior)	●
HIPPOCAMPAL REGION	
Entorhinal cortex Layer II	● ●
Entorhinal cortex Layer III	● ●
Entorhinal cortex Layer IV	● ●
Entorhinal cortex Layers V-VI	● ●
Subiculum (deep)	● ●
Subiculum (superficial)	●
Ammon's Horn CA3	●
OLFACTORY BULB	
Mitral cells (main bulb)	● ●
Internal tufted cells (main bulb)	● ●
External tufted cells (main bulb)	●
ANTERIOR OLFACTORY NUCLEUS	
Pars externa	● ●
AON proper	●

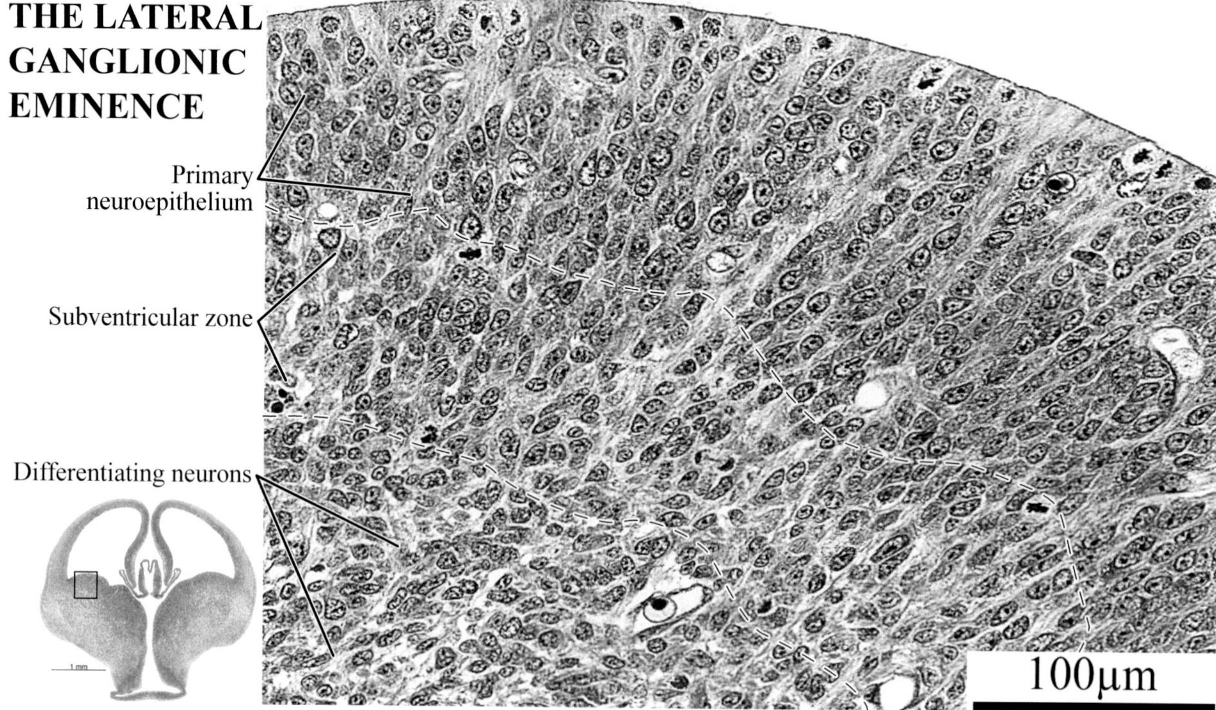
THE LATERAL GANGLIONIC EMINENCE
Primary neuroepithelium
Subventricular zone
Differentiating neurons
100µm

Figure 14. Frontal section through the telencephalon showing the neuroepithelium and subventricular zone of the lateral ganglionic eminence in an E16 rat embryo at the same developmental stage as the human specimens in this volume. (3 µ methacrylate section, toluidine blue stain)
Source: braindevelopmentmaps.org (E16 coronal archive)

TELENCEPHALON

There is still robust neurogenesis throughout the basal telencephalon (**Table 1E**). Many neurons have already been generated and the parenchyma is expanding. One region where neurons will continue to be generated is in the lateral (**Fig. 14**) and medial (**Fig. 15**) ganglionic eminences. The enormous subventricular zone is becoming established and mitotic cells are scattered outside the basal neuroepithelium. The amygdala and septum, also still busy with the final bout of neurogenesis are expanding (not illustrated) and neurons are being generated in nearly every neuronal population.

Neurogenesis in the cortical regions (**Table 1F**) of the telencephalon are producing the latest-generated Cajal-Retzius neurons and subplate neurons. These cells are obvious in both dorsomedial (**Fig. 16A**) and ventrolateral (**Fig. 16B**) areas. The most notable feature of neo-

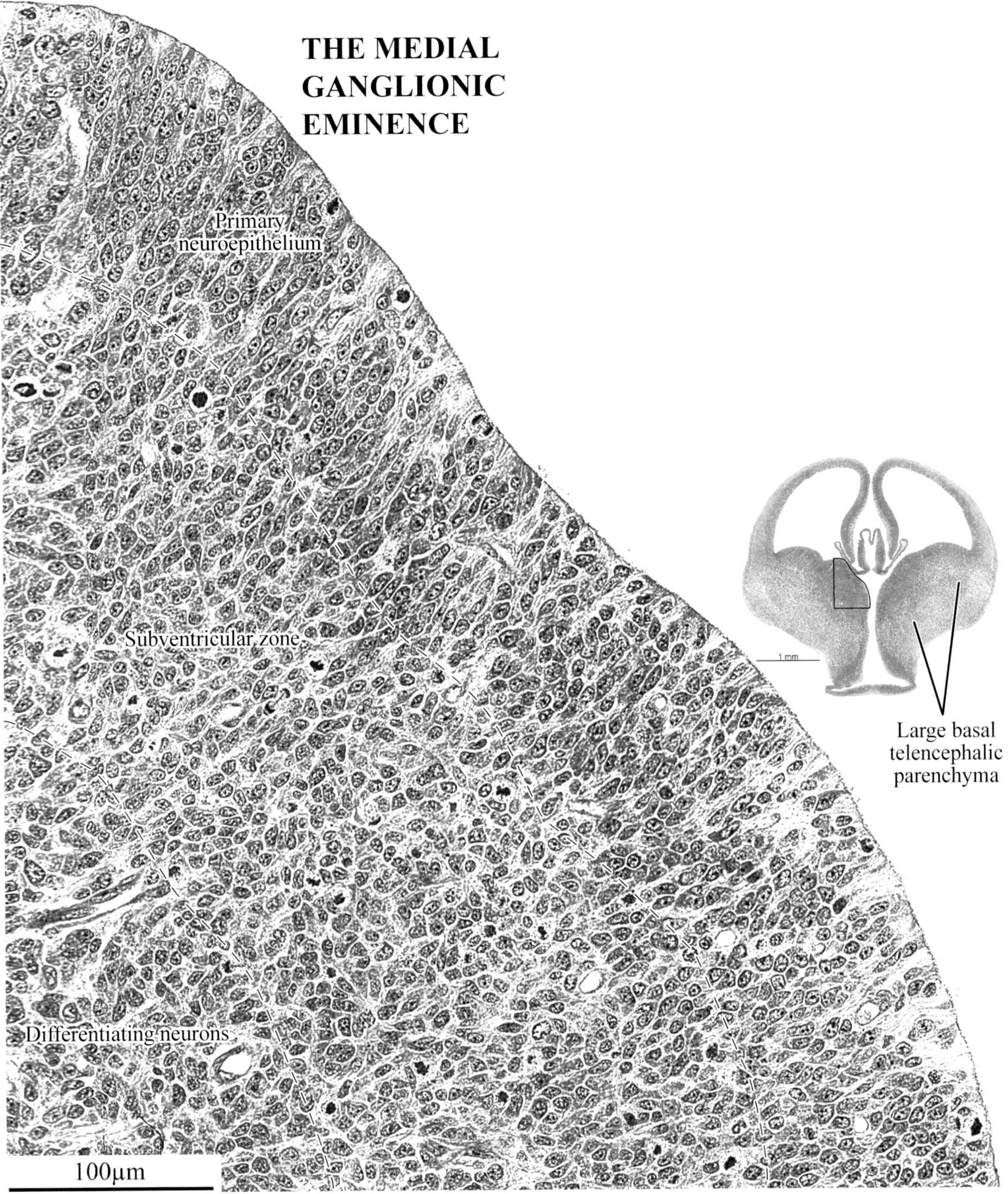
THE MEDIAL GANGLIONIC EMINENCE
Primary neuroepithelium
Subventricular zone
Differentiating neurons
100μm
1 mm
Large basal telencephalic parenchyma

Figure 15. Frontal section through the telencephalon showing the neuroepithelium and subventricular zone of the medial ganglionic eminence in an E16 rat embryo at the same developmental stage as the human specimens in this volume. (3 μ methacrylate section, toluidine blue stain)
Source: braindevelopmentmaps.org (E16 coronal archive)

THE CEREBRAL CORTEX

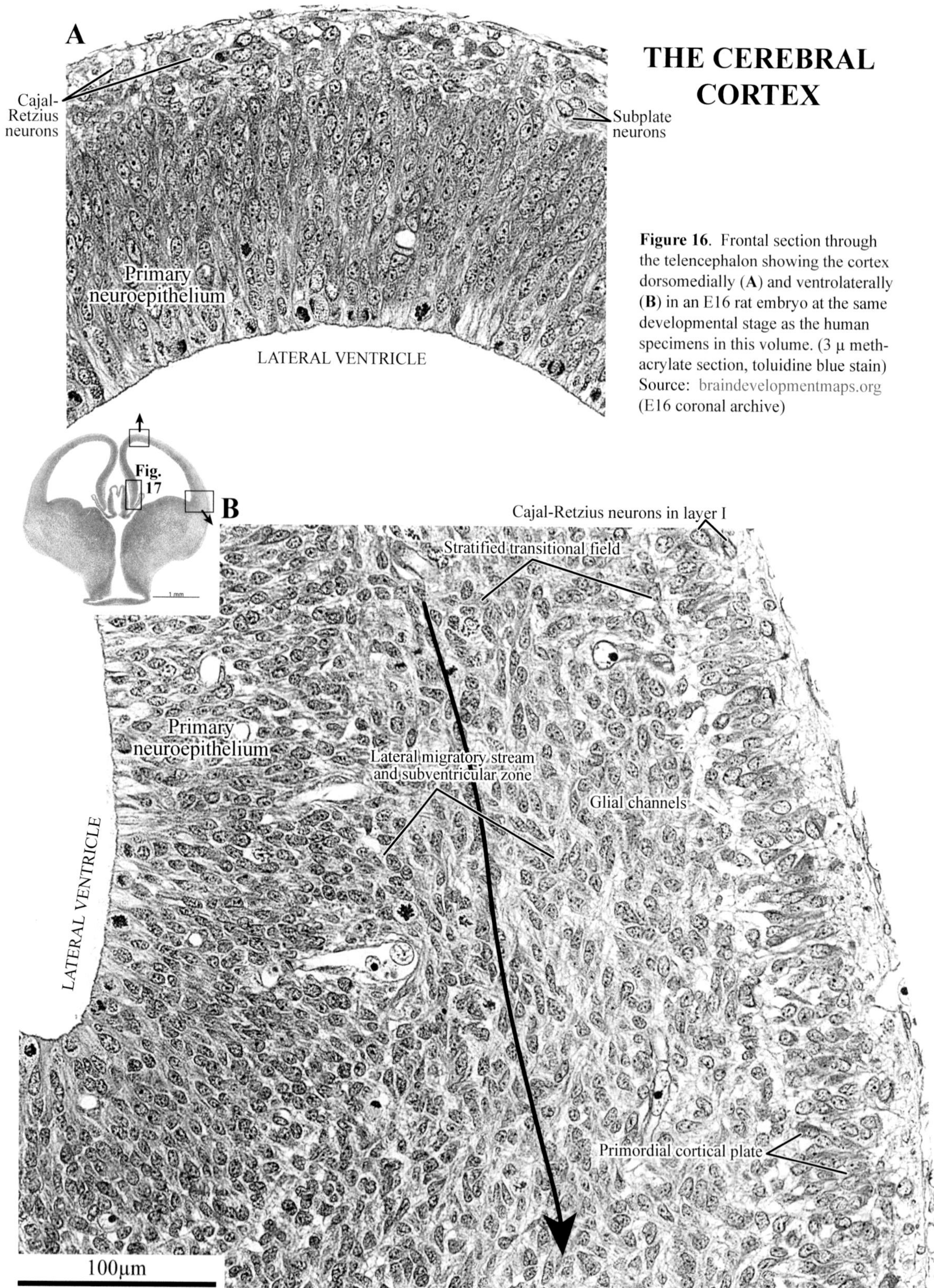
A
Cajal-Retzius neurons
Subplate neurons
Primary neuroepithelium
LATERAL VENTRICLE
Fig. 17
B
Cajal-Retzius neurons in layer I
Stratified transitional field
Primary neuroepithelium
Lateral migratory stream and subventricular zone
Glial channels
LATERAL VENTRICLE
Primordial cortical plate
100μm

Figure 16. Frontal section through the telencephalon showing the cortex dorsomedially (**A**) and ventrolaterally (**B**) in an E16 rat embryo at the same developmental stage as the human specimens in this volume. (3 μ methacrylate section, toluidine blue stain) Source: braindevelopmentmaps.org (E16 coronal archive)

cortical development is the first appearance of the cortical plate in the earlier-maturing ventrolateral region which will become the insular cortex. Our autoradiographic studies in rats (Bayer and Altman, 1991) indicate that the first cells in the cortical plate are likely subplate neurons that transiently take on a radial orientation. Later, as layer VI neurons migrate toward Layer I and settle above subplate neurons, the original occupants of the cortical plate delaminate and become polymorphic. The white spaces beneath the primordial cortical plate are characteristic of every specimen in this age group and may be glial channels for axons growing into the neocortex. Early-generated neurons that will settle in Layers V and VI are migrating in the stratified transitional field, (especially ventrolaterally), yet the cortical neuroepithelium is still generating neurons for these layers. Neurons in layers IV-II are just getting started.

Table 1F shows that every layer in the entorhinal cortex is being generated, continuing the trend started in the previous age group (Bayer and Altman, 2022, Volume 3, Table 1G). Deep neurons in the subiculum continue their neurogenesis and now some superficial cells are being added. A few neurons in the CA3 field of hippocampus proper are originating. All other neuronal populations have not yet started. **Figure 17** shows large neurons beneath the pia in the curved hippocampal primordium. Our studies indicate that these are the neurons in the superficial stratum lacunosum moleculare and the deep stratum oriens, a modified primordial plexiform layer.

Deep neurons throughout the olfactory cortex started neurogenesis in human specimens with 10-mm crown rump lengths (Bayer and Altman, 2023, Volume 2, Table 1E) and all layers are continuing neurogenesis in the specimens in this volume (**Table 1F**). **Figure 18** shows the lamination in the olfactory cortex, especially in layer II. Layers III-IV contain many spindle-shaped cells and we postulate that these are neurons produced in the cortical neuroepithelium that migrate (via the lateral migratory stream) into the olfactory cortex.

Figure 19 is a sagittal section through the olfactory bulb, olfactory nerve entry zone, and the olfactory epithelium in an E16 rat embryo at a similar stage as the human specimens in this volume. The olfactory nerve bundles envelop the evaginating bulb, and the pial membrane is completely obliterated. Just behind the olfactory evagination are large cells with copious cytoplasm leaving the olfactory epithelium. There are no barriers to these cells entering the brain, and they may be the gonadotropin-releasing hormone neurons (GNRH) that migrate in with the olfactory nerve (Wray et al., 1989). There are many spindle-shaped cells in the olfactory bulb itself that are migrating into the olfactory bulb from the basal telencephalon. Mitral cells are produced in the basal telencephalic neuroepithelium and migrate into the bulb (Bayer, 2017). Some mitral neurons are generated in human specimens with 10-mm crown rump lengths (Bayer and Altman, 2022, Volume 2, Table 1E) and some of those early-generated neurons may have already reached the bulb in the specimens in this volume. The small inset in **Figure 19** shows fascicles of the vomeronasal nerve which will surround the accessory olfactory bulb, where the output neurons are among the oldest in the olfactory sensory system.

THE HIPPOCAMPUS

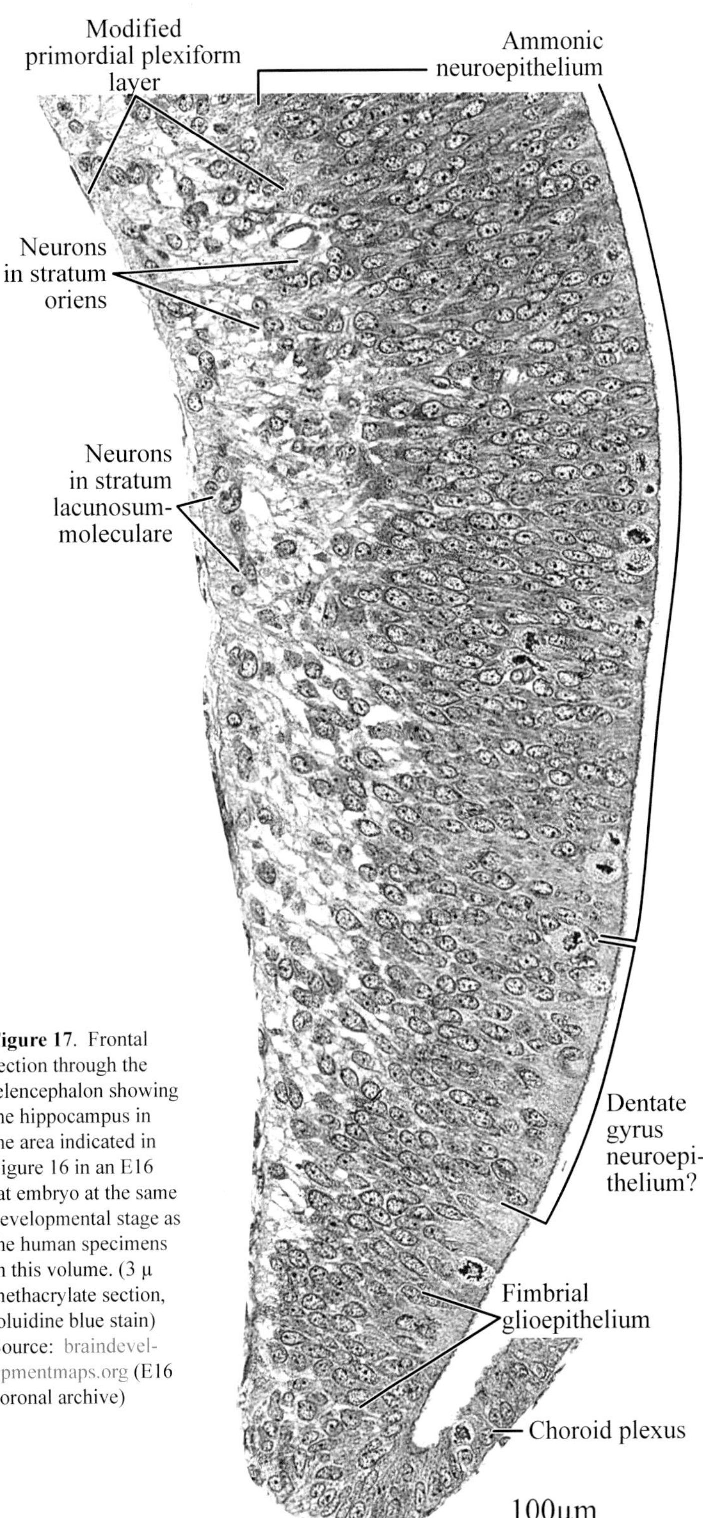
Modified primordial plexiform layer
Ammonic neuroepithelium
Neurons in stratum oriens
Neurons in stratum lacunosum-moleculare
Dentate gyrus neuroepithelium?
Fimbrial glioepithelium
Choroid plexus
100μm

Figure 17. Frontal section through the telencephalon showing the hippocampus in the area indicated in Figure 16 in an E16 rat embryo at the same developmental stage as the human specimens in this volume. (3 μ methacrylate section, toluidine blue stain) Source: braindevelopmentmaps.org (E16 coronal archive)

THE OLFACTORY CORTEX

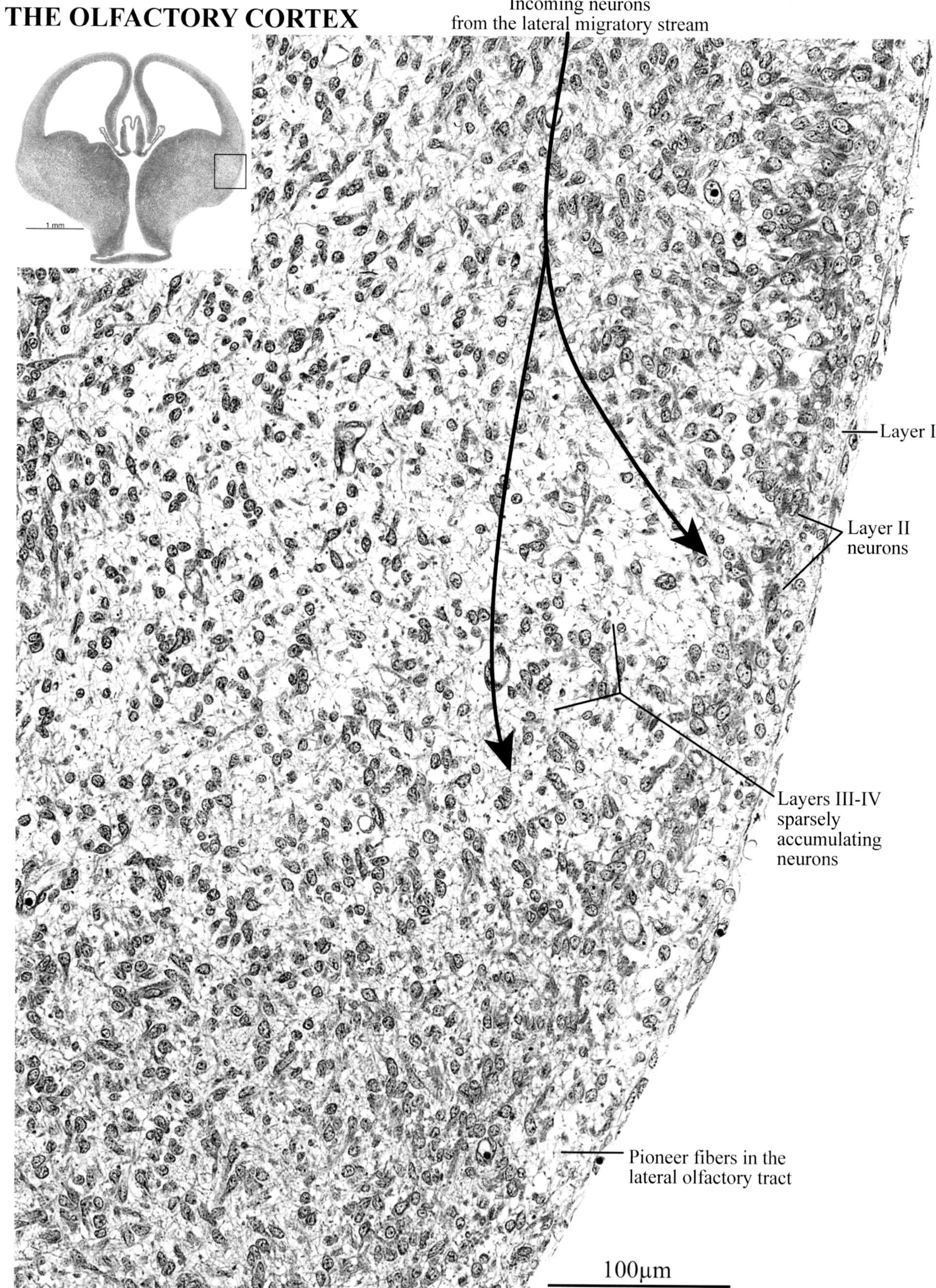
Incoming neurons
from the lateral migratory stream
1 mm
Layer I
Layer II
neurons
Layers III-IV
sparsely
accumulating
neurons
Pioneer fibers in the
lateral olfactory tract
100μm

Figure 18. Frontal section through the telencephalon showing the olfactory cortex in an E16 rat embryo at the same developmental stage as the human specimens in this volume. (3 μ methacrylate section, toluidine blue stain) braindevelopmentmaps.org (E16 coronal archive)

THE OLFACTORY BULB AND OLFACTORY EPITHELIUM

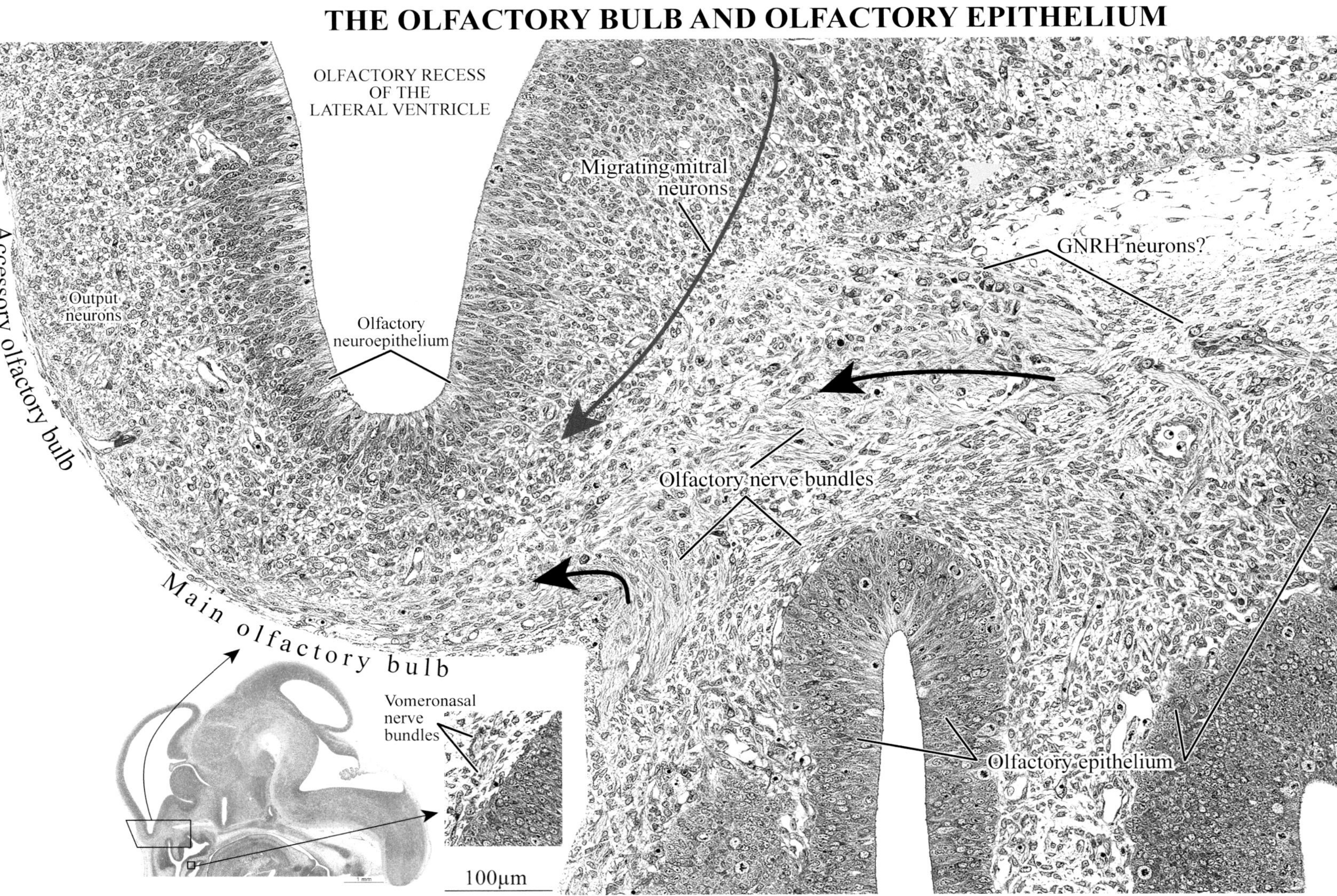
OLFACTORY RECESS
OF THE
LATERAL VENTRICLE
Migrating mitral
neurons
GNRH neurons?
Output
neurons
Olfactory
neuroepithelium
Accessory olfactory bulb
Olfactory nerve bundles
Main olfactory bulb
Vomeronasal
nerve
bundles
Olfactory epithelium
100µm

Figure 19. Sagittal section through the telencephalon showing the olfactory bulb and olfactory epithelium in an E16 rat embryo at the same developmental stage as the human specimens in this volume. *Red arrow*, trajectory of migrating mitral cells; *yellow outlines and arrows*, presumptive GNRH neurons migrating into the brain, *large black arrows*, olfactory nerve bundles. (3 µ methacrylate section, toluidine blue stain) braindevelopmentmaps.org (E16 sagittal archive)

REFERENCES

Altman J, Bayer SA (1984) *The Development of the Rat Spinal Cord.* Advances in Anatomy Embryology and Cell Biology, Vol. 85, Berlin, Springer-Verlag.

Bayer, SA (2013-present) www.braindevelopmentmaps.org (This website is an image database of methacrylate-embedded normal rat embryos and paraffin-embedded rat embryos exposed to ^{3}H-Thymidine.)

Bayer SA (2017) *Development of the Olfactory Bulb.* braindevelopmentmaps.org. Laboratory of Developmental Neurobiology, Ocala, FL.

Bayer, SA (2022) *Glossary to Accompany Atlas of Human Central Nervous System Development (Second Edition)* Laboratory of Developmental Neurobiology, Ocala, FL.

Bayer SA, Altman J (1991) *Neocortical Development*, Raven Press, New York.

Bayer SA, Altman J, Russo RJ, Zhang X (1993) Timetables of neurogenesis in the human brain based on experimentally determined patterns in the rat. *Neurotoxicology* **14**: 83-144.

Bayer SA, Altman J, Russo RJ, Zhang X (1995) Embryology. In: *Pediatric Neuropathology*, Serge Duckett, Ed. Williams and Wilkins, pp. 54-107.

Bayer SA, Altman J (1995) Development: Some principles of neurogenesis, neuronal migration and neural circuit formation. In: *The Rat Nervous System*, 2nd Edition, George Paxinos, Ed. Academic Press, Orlando, Florida., pp. 1079-1098.

Bayer SA, Altman J (2006) *Atlas of Human Central Nervous System Development* (First Edition), Volume 4, CRC Press.

Bayer SA, Altman J (2012-present) www.neurondevelopment.org (This website has downloadable pdf files of our scientific papers on rat brain development grouped by subject.)

Bayer SA, Altman J (2022) *The Human Brain during the First Trimester 6.3- to 10.5-mm Crown-Rump Lengths, Atlas of Human Central Nervous System Development* (Second Edition), Volume 2, CRC Press/Taylor and Francis.

Bayer SA, Altman J (2022) *The Human Brain during the First Trimester 16- to 18-mm Crown-Rump Lengths, Atlas of Human Central Nervous System Development* (Second Edition), Volume 3, CRC Press/Taylor and Francis.

Loughna P, Citty L, Evans T, Chudleigh T (2009) Fetal size and dating: Charts recommended for clinical obstetric practice, *Ultrasound*, 17:161-167.

O'Rahilly R, Müller F. (1987) *Developmental Stages in Human Embryos, Carnegie Institution of Washington*, Publication 637.

O'Rahilly R; Müller F. (1994) *The Embryonic Human Brain*, Wiley-Liss, New York.

Van Hartesveldt C, Moore B, Hartman, BK (1986) Transient midline raphe glial structure in the developing rat *J Comp Neurol* 253:174-84.

Wray S, Grant P, Gainer, H (1989) Evidence that cells expressing luteinizing hormone-releasing hormone mRNA in the mouse are derived from progenitor cells in the olfactory placode. *Proc. Natl Acad. Sci. USA*, 86:8132-8136.

PART II: C6202 CR 21 mm (GW 8.0) Sagittal

This is specimen 6202 in the Carnegie Collection, designated here as C6202, is a fetus of unknown sex with a crown rump length (CR) of 21 mm estimated to be at gestational week (GW) 8.0. The entire fetus was cut in the sagittal plane in 20-μm sections and stained with hematoxylin and eosin. Information on the date of specimen collection, fixative, and embedding medium was not available to us. Since there is no photograph of this specimen before histological processing, a specimen from Hochstetter (1919) that is comparable in age to C6202 is used to show external brain features at GW8.0 (**A**, **Figure 20**). Like most sagittal specimens, C6202's sections are not cut parallel to the midline; **Figure 20** shows the approximate rotations in horizontal (**B**) and vertical (**C**) dimensions. Photographs of 7 sections (**Levels 1-7**) are illustrated at low magnification in four-parts (**Plates 1-7A-D**). The **A/B** parts show the brain in place in the skull; the **C/D** parts show only the brain (and some peripheral ganglia) at slightly higher magnification. **Plates 8-15** show high-magnification views of various parts of the brain at different levels from the cerebral cortex (**Plate 8**) to the midline raphe glial structure in the midbrain and cervical spinal cord (**Plate 15**). Most of the high-magnification plates are rotated 90° (landscape orientation) to more efficiently use page space.

The ***superventricles*** are large in the centers of all brain structures, especially in the telencephalon and rhombencephalon. Sections near the midline show the enormous size of the diencephalic and mesencephalic superventricles. The respective thicknesses of the ***neuroepithelium*** (NEP) and parenchyma are keys to determining the degree of maturation of various brain structures.

The parenchyma is thick and bordered by a thin NEP that is transitioning to the ependymal layer in the medial medulla and pons. Neurogenesis of mostly motor structures is complete here but gliogenesis is still robust. The one exception is that midline raphe neurons in the pons have robust neurogenesis at this time. The parenchyma is thinner and the NEP is thicker in the lateral medulla and pons where mainly sensory neural populations are still being generated. The production of late-generated neurons is beginning to reach a peak in the precerebellar NEP (pontine nuclei and reticular pontine nucleus). Many neurons are in the *posterior intramural migratory stream* (not easily visible in sagittal sections), and some have accumulated in the inferior olive in the posteromedial medulla (**Plate 1D**).

The cerebellar parenchyma has multiple layers in the ***cerebellar transitional field (CTF,* Plate 13*)***. The cells in ***CTF2*** and ***3*** are probably deep neurons that will eventually settle in the dentate, interpositus, and fastigial nuclei. ***CTF4*** is an intermingling of unidentified cells and fibers. The cerebellar NEP is thick but most Purkinje cells have already been generated. The slightly less dense basal parts of the NEP probably contain the large population of sequestered postmitotic Purkinje cells. The NEP itself may be generating progenitors of Golgi cells that will disperse among the future granule cell layer of the cerebellar cortex, which has yet to develop.

There are layers of dense cells adjacent to the mesencephalic tegmental NEP that has thinned considerably compared to younger specimens. Many waves of neurons are sojourning and intermingling with incoming fibers from the spinal cord and medulla (**Plate 12**). Nuclear groupings in the tegmentum will be easier to see in the next frontal/horizontal specimen, C966. The mesencephalic tectal NEP is very thick and lies adjacent to a thin parenchyma. Many neurons in both the superior and inferior colliculi are currently being generated in these NEPs, so it is still too early to see cells migrating out.

The diencephalic NEP is thick and the adjacent parenchyma is filled with dense zones of sojourning and migrating neurons, especially in the hypothalamus. The subthalamus stands out as the most mature diencephalic area.

The basal telencephalic and basal ganglionic NEPs are thick, and the oldest neurons (e.g., globus pallidus) are settling in the adjacent parenchyma. Many neurons are currently being generated in the basal telencephalon. In the cerebral cortex, the NEP is bordered by a thin primordial plexiform layer that contains the oldest cortical neurons (Cajal-Retzius cells) and subplate neurons. The cortical NEP is expanding and increasing its number of neural stem cells as the telencephalic superventricle grows; some neurons in layers V-VI are currently being generated in the cortical NEPs and nearly all cortical neurons in layers II-IV have not yet been born.

The most immature part of the cerebral cortex is Ammon's horn in the hippocampus. The hippocampal NEP is the only one in stockbuilding stage throughout the entire brain.

GW8.0 SAGITTAL

A perfect sagittal cut through the brain bisects the cerebral cortex into two separate hemispheres by passing through the interhemispheric fissure, and does the same in the brainstem by passing through the midline of the ventricles.

Sections of C6202's brain rotate 10.5° counterclockwise from the horizontal midline running through the cerebral cortex and midbrain tectum (top view). C6202's sections are quite close to the vertical midline, rotating only 2.8° counterclockwise (back view). In the sections illustrated on the following pages, the telencephalon and diencephalon (top) are tilted away from the observer, while the medulla and upper spinal cord (bottom) are tilted toward the observer.

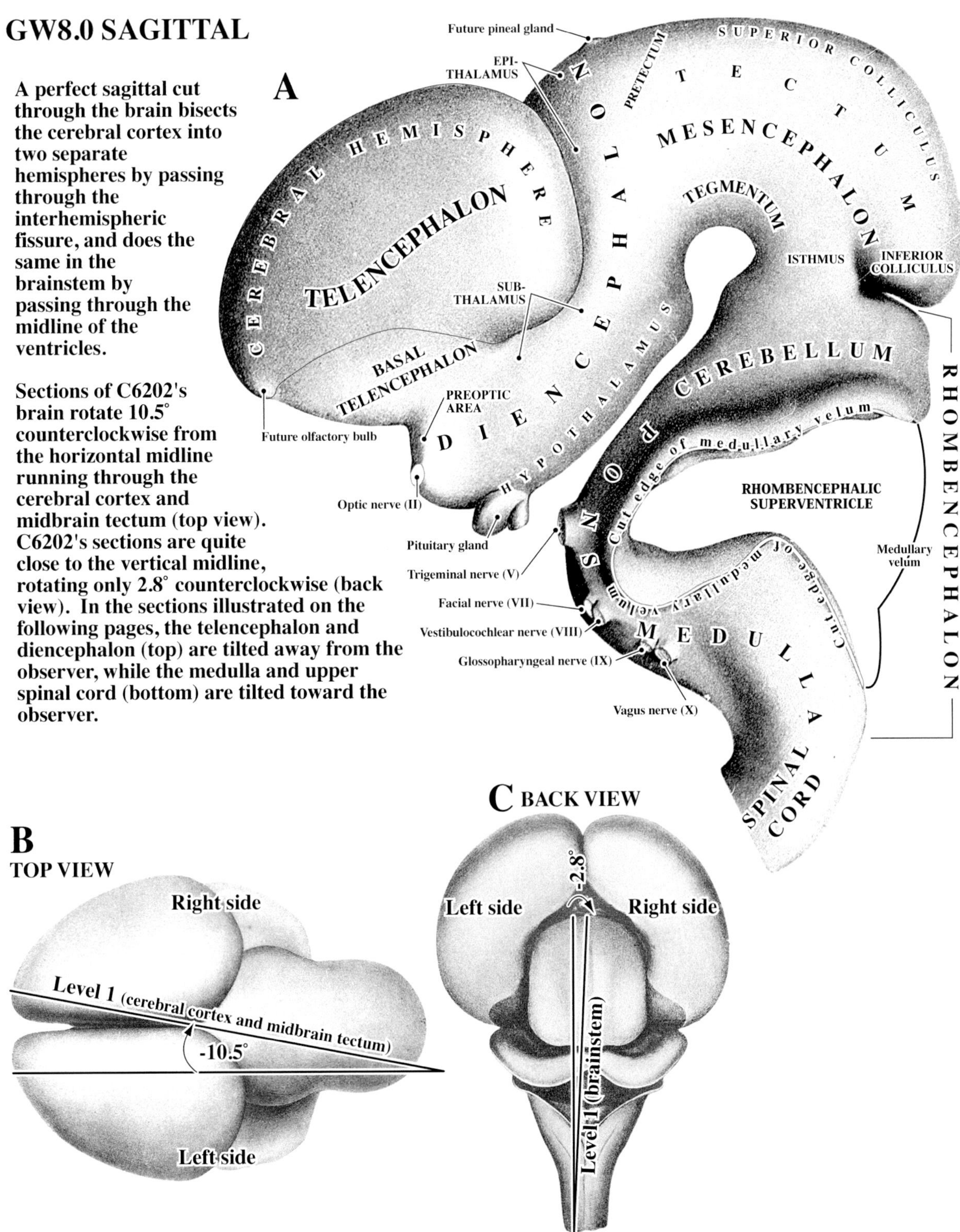
A
Future pineal gland
EPI-THALAMUS
SUPERIOR COLLICULUS
PRETECTUM
TECTUM
MESENCEPHALON
TEGMENTUM
ISTHMUS
INFERIOR COLLICULUS
CEREBRAL HEMISPHERE
TELENCEPHALON
SUB-THALAMUS
BASAL TELENCEPHALON
PREOPTIC AREA
DIENCEPHALON
HYPOTHALAMUS
Future olfactory bulb
Optic nerve (II)
Pituitary gland
CEREBELLUM
PONS
Cut edge of medullary velum
RHOMBENCEPHALIC SUPERVENTRICLE
RHOMBENCEPHALON
Medullary velum
Trigeminal nerve (V)
Facial nerve (VII)
Vestibulocochlear nerve (VIII)
Glossopharyngeal nerve (IX)
Vagus nerve (X)
MEDULLA
SPINAL CORD
B
TOP VIEW
Right side
Level 1 (cerebral cortex and midbrain tectum)
-10.5°
Left side
C BACK VIEW
Left side
Right side
-2.8°
Level 1 (brainstem)

Figure 20. **A**, the lateral view of the brain and upper cervical spinal cord from a specimen with a crown rump length of 19.4 mm (modified from Figure 35, Table VI, Hochstetter, 1919) identifies external features of a brain similar to C6202 (CR 21 mm). **B**, top view of the brain in **A** (modified from Figure 37, Table VI, Hochstetter, 1919) shows how C6202's sections rotate from a line parallel to the horizontal midline in the interhemispheric fissure and midbrain tectum. **C**, back view of the more mature brain with a crown rump length of 38 mm (modified from Figure 44, Table VIII, Hochstetter, 1919) shows how C6202's sections rotate from a line parallel to the vertical midline in the brainstem and upper cervical spinal cord.

PLATE 1A

CR 21 mm, GW 8, C6202
Sagittal, Slide 30, Section 2
SKULL, MAJOR BRAIN STRUCTURES, AND VENTRICULAR DIVISIONS

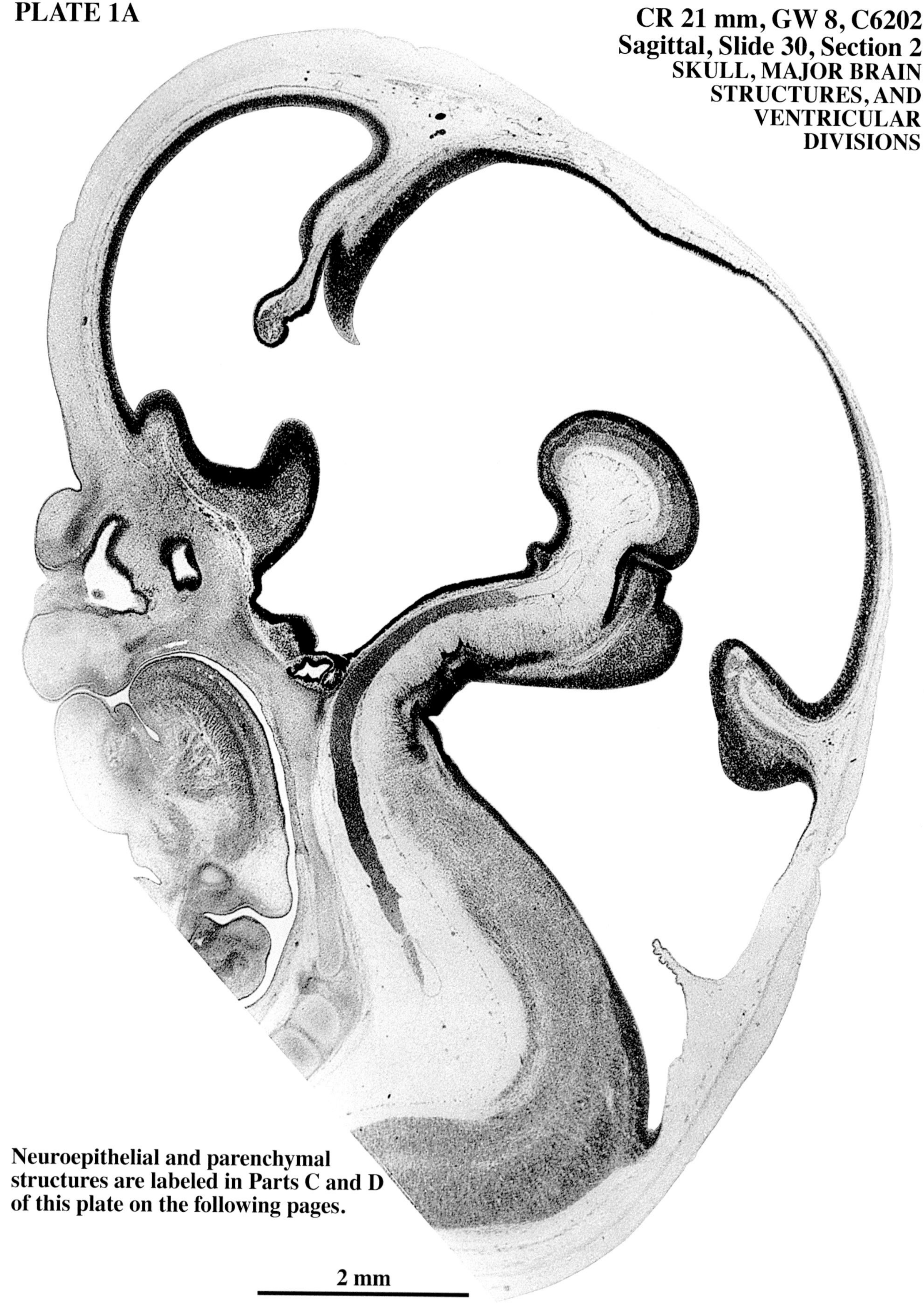

2 mm

Neuroepithelial and parenchymal structures are labeled in Parts C and D of this plate on the following pages.

PLATE 1B

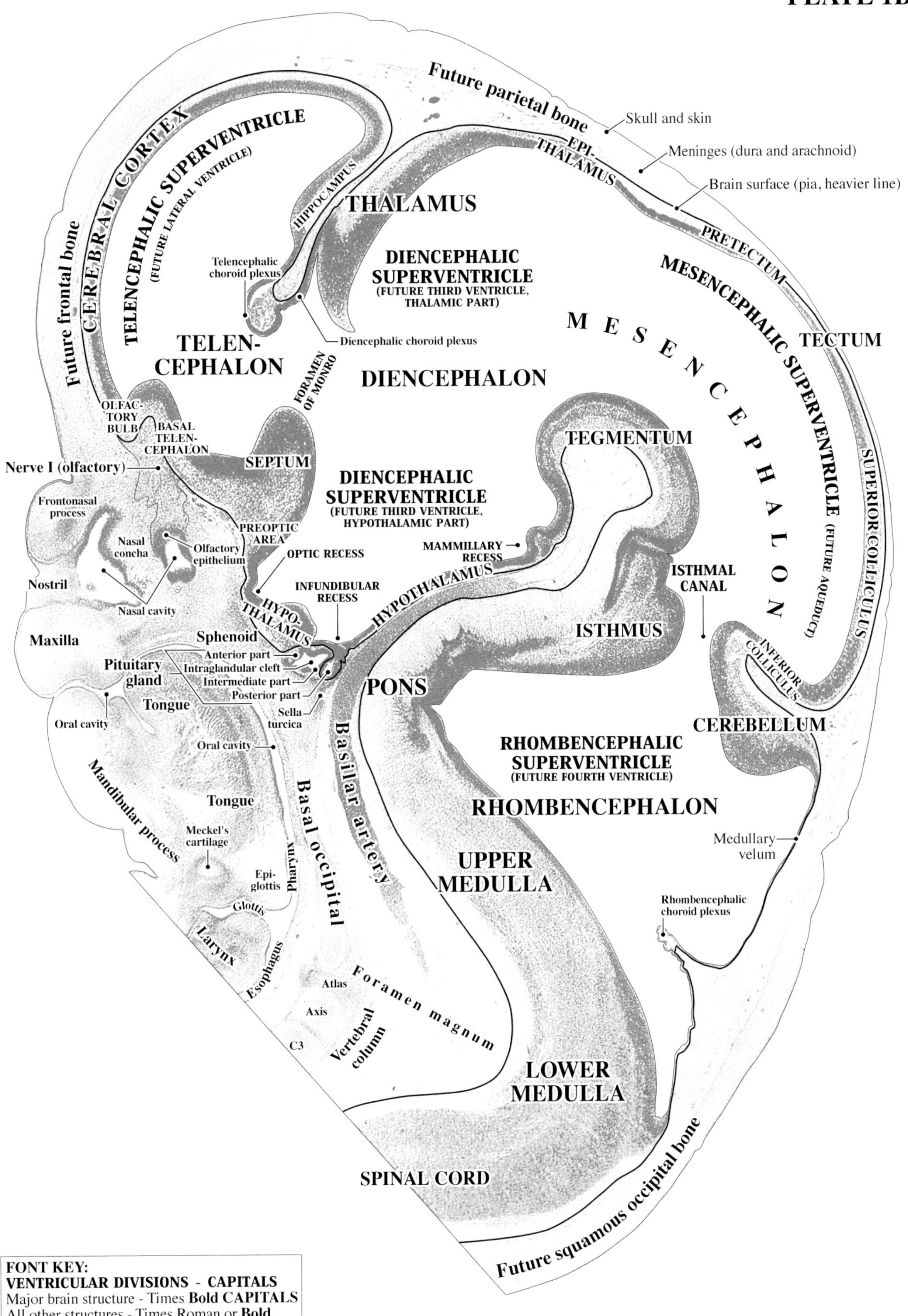
Future parietal bone
Skull and skin
Meninges (dura and arachnoid)
Brain surface (pia, heavier line)
CEREBRAL CORTEX
TELENCEPHALIC SUPERVENTRICLE
(FUTURE LATERAL VENTRICLE)
HIPPOCAMPUS
THALAMUS
EPI-
THALAMUS
PRETECTUM
Future frontal bone
Telencephalic choroid plexus
DIENCEPHALIC SUPERVENTRICLE
(FUTURE THIRD VENTRICLE, THALAMIC PART)
MESENCEPHALIC SUPERVENTRICLE
(FUTURE AQUEDUCT)
TECTUM
M E S E N C E P H A L O N
Diencephalic choroid plexus
TELEN-
CEPHALON
FORAMEN OF MONRO
DIENCEPHALON
OLFAC-
TORY
BULB
BASAL
TELEN-
CEPHALON
SEPTUM
TEGMENTUM
SUPERIOR COLLICULUS
Nerve I (olfactory)
Frontonasal process
DIENCEPHALIC SUPERVENTRICLE
(FUTURE THIRD VENTRICLE, HYPOTHALAMIC PART)
PREOPTIC AREA
Nasal concha
Olfactory epithelium
OPTIC RECESS
MAMMILLARY RECESS
ISTHMAL CANAL
Nostril
Nasal cavity
INFUNDIBULAR RECESS
HYPOTHALAMUS
HYPO-
THALAMUS
ISTHMUS
Maxilla
Sphenoid
Anterior part
Intraglandular cleft
Intermediate part
Posterior part
Pituitary gland
INFERIOR COLLICULUS
PONS
Tongue
Oral cavity
Sella turcica
CEREBELLUM
Oral cavity
RHOMBENCEPHALIC SUPERVENTRICLE
(FUTURE FOURTH VENTRICLE)
Mandibular process
Basilar artery
Basal occipital
Tongue
RHOMBENCEPHALON
Meckel's cartilage
Medullary velum
UPPER MEDULLA
Epi-
glottis
Pharynx
Glottis
Rhombencephalic choroid plexus
Larynx
Esophagus
Atlas
Foramen magnum
Axis
C3
Vertebral column
LOWER MEDULLA
SPINAL CORD
Future squamous occipital bone
FONT KEY:
VENTRICULAR DIVISIONS - CAPITALS
Major brain structure - Times Bold CAPITALS
All other structures - Times Roman or Bold

PLATE 1C

CR 21 mm, GW 8, C6202
Sagittal, Slide 30, Section 2
NEUROEPITHELIAL AND PARENCHYMAL BRAIN STRUCTURES

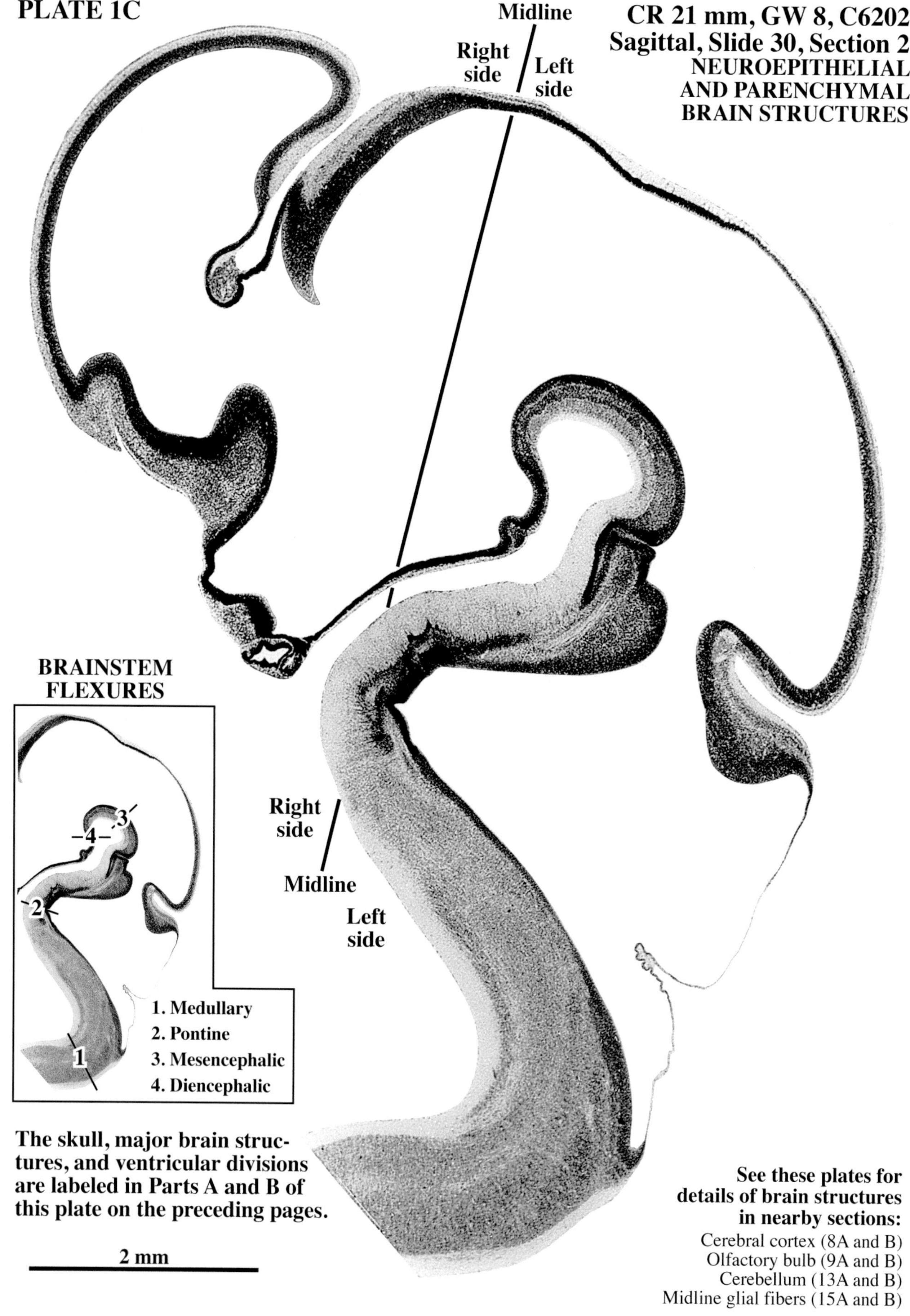

Midline
Right side
Left side
Right side
Midline
Left side
BRAINSTEM FLEXURES
3
4
2
1
1. Medullary
2. Pontine
3. Mesencephalic
4. Diencephalic
2 mm

The skull, major brain structures, and ventricular divisions are labeled in Parts A and B of this plate on the preceding pages.

See these plates for details of brain structures in nearby sections:
Cerebral cortex (8A and B)
Olfactory bulb (9A and B)
Cerebellum (13A and B)
Midline glial fibers (15A and B)

PLATE 1D

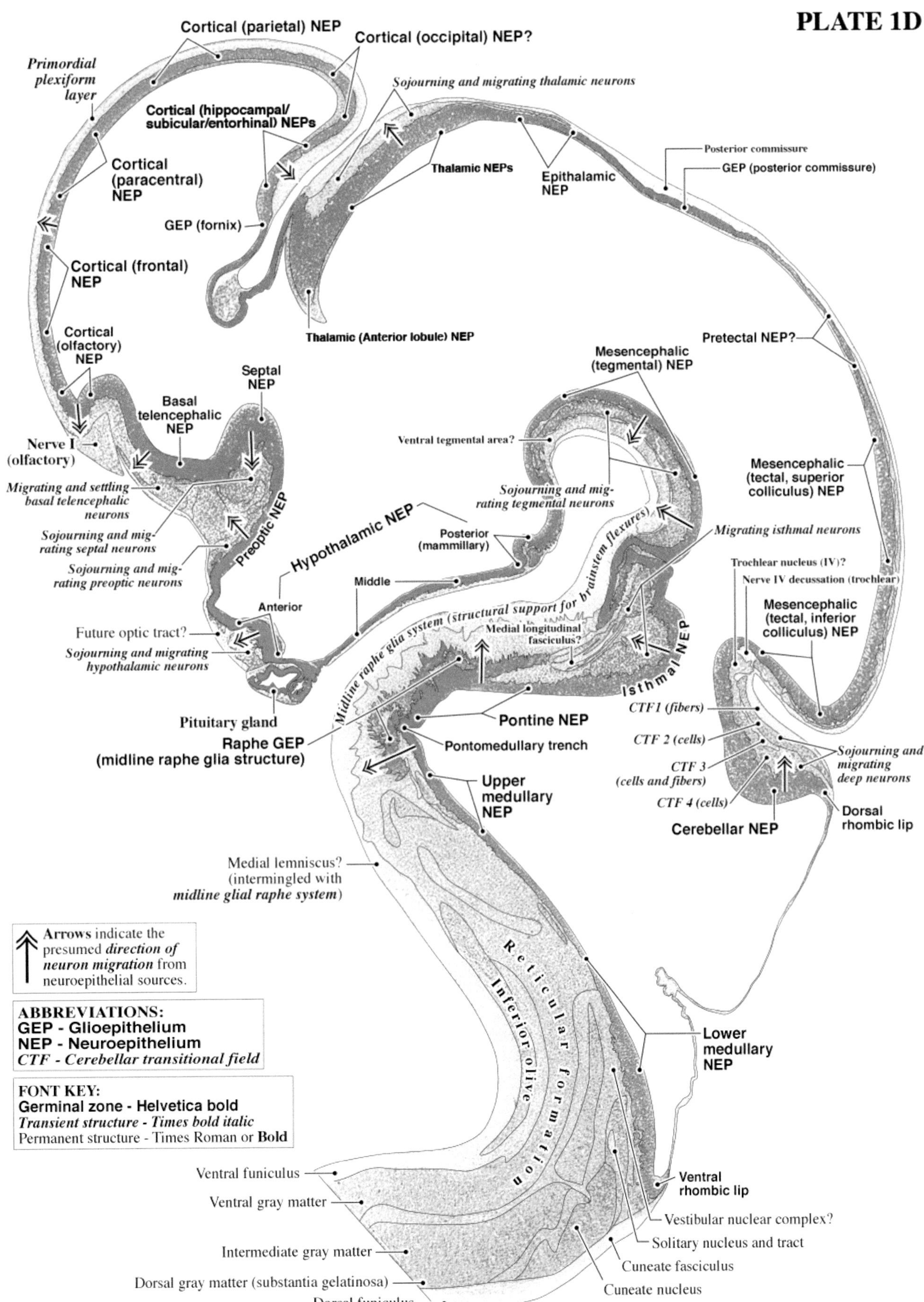

Cortical (parietal) NEP
Cortical (occipital) NEP?
Primordial plexiform layer
Sojourning and migrating thalamic neurons
Cortical (hippocampal/ subicular/entorhinal) NEPs
Cortical (paracentral) NEP
Thalamic NEPs
Epithalamic NEP
Posterior commissure
GEP (posterior commissure)
GEP (fornix)
Cortical (frontal) NEP
Cortical (olfactory) NEP
Thalamic (Anterior lobule) NEP
Pretectal NEP?
Mesencephalic (tegmental) NEP
Septal NEP
Basal telencephalic NEP
Nerve I (olfactory)
Ventral tegmental area?
Mesencephalic (tectal, superior colliculus) NEP
Migrating and settling basal telencephalic neurons
Sojourning and migrating tegmental neurons
Preoptic NEP
Hypothalamic NEP
Posterior (mammillary)
Migrating isthmal neurons
Sojourning and migrating septal neurons
Sojourning and migrating preoptic neurons
Trochlear nucleus (IV)?
Nerve IV decussation (trochlear)
Middle
Anterior
Midline raphe glia system (structural support for brainstem flexures)
Medial longitudinal fasciculus?
Mesencephalic (tectal, inferior colliculus) NEP
Future optic tract?
Sojourning and migrating hypothalamic neurons
Isthmal NEP
Pituitary gland
CTF1 (fibers)
Pontine NEP
Raphe GEP (midline raphe glia structure)
Pontomedullary trench
CTF 2 (cells)
CTF 3 (cells and fibers)
Sojourning and migrating deep neurons
CTF 4 (cells)
Upper medullary NEP
Dorsal rhombic lip
Cerebellar NEP
Medial lemniscus? (intermingled with midline glial raphe system)
Arrows indicate the presumed direction of neuron migration from neuroepithelial sources.
Reticular formation
Inferior olive
ABBREVIATIONS:
GEP - Glioepithelium
NEP - Neuroepithelium
CTF - Cerebellar transitional field
Lower medullary NEP
FONT KEY:
Germinal zone - Helvetica bold
Transient structure - Times bold italic
Permanent structure - Times Roman or Bold
Ventral funiculus
Ventral rhombic lip
Ventral gray matter
Vestibular nuclear complex?
Solitary nucleus and tract
Intermediate gray matter
Cuneate fasciculus
Dorsal gray matter (substantia gelatinosa)
Cuneate nucleus
Dorsal funiculus

PLATE 2A

CR 21 mm, GW 8, C6202
Sagittal, Slide 26, Section 2
SKULL, MAJOR BRAIN STRUCTURES, AND VENTRICULAR DIVISIONS

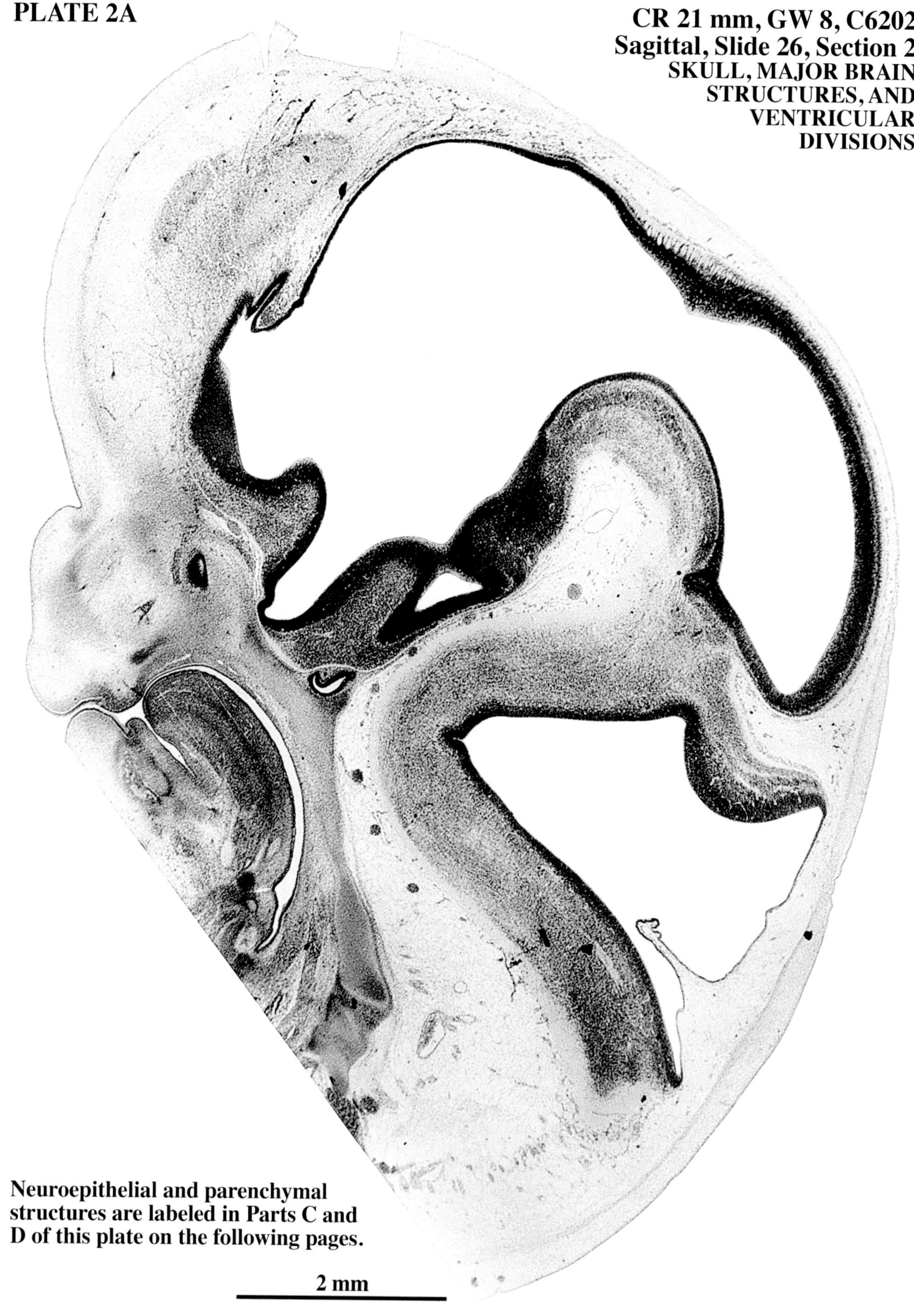

Neuroepithelial and parenchymal structures are labeled in Parts C and D of this plate on the following pages.

2 mm

PLATE 2B

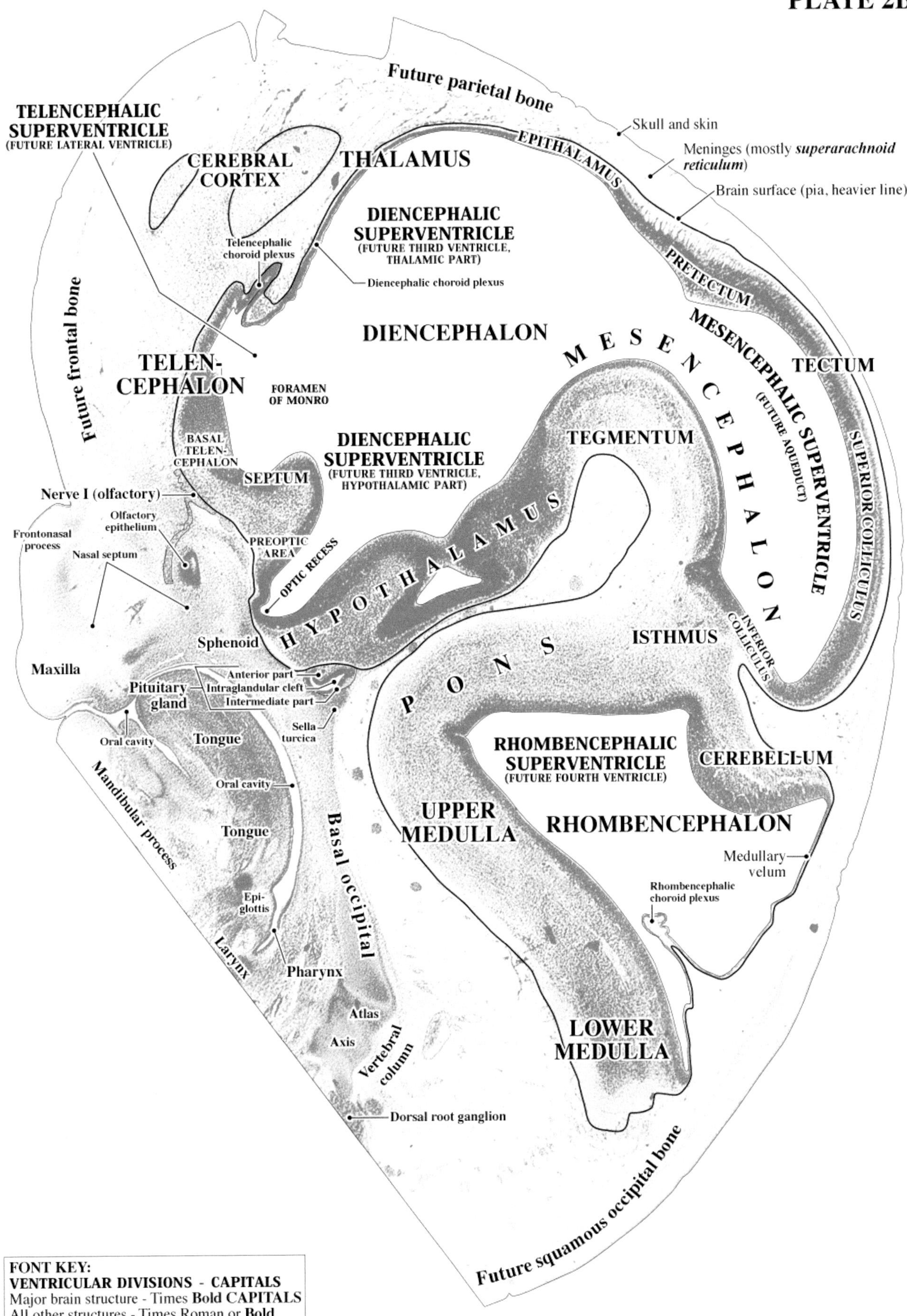
Future parietal bone
Skull and skin
Meninges (mostly *superarachnoid reticulum*)
Brain surface (pia, heavier line)
TELENCEPHALIC SUPERVENTRICLE
(FUTURE LATERAL VENTRICLE)
CEREBRAL CORTEX
THALAMUS
EPITHALAMUS
DIENCEPHALIC SUPERVENTRICLE
(FUTURE THIRD VENTRICLE, THALAMIC PART)
Telencephalic choroid plexus
Diencephalic choroid plexus
PRETECTUM
Future frontal bone
DIENCEPHALON
MESENCEPHALIC SUPERVENTRICLE
(FUTURE AQUEDUCT)
TECTUM
TELEN-CEPHALON
FORAMEN OF MONRO
MESENCEPHALON
SUPERIOR COLLICULUS
BASAL TELEN-CEPHALON
DIENCEPHALIC SUPERVENTRICLE
(FUTURE THIRD VENTRICLE, HYPOTHALAMIC PART)
TEGMENTUM
SEPTUM
Nerve I (olfactory)
Olfactory epithelium
Frontonasal process
Nasal septum
PREOPTIC AREA
OPTIC RECESS
HYPOTHALAMUS
Sphenoid
ISTHMUS
INFERIOR COLLICULUS
PONS
Maxilla
Anterior part
Pituitary gland
Intraglandular cleft
Intermediate part
Sella turcica
Oral cavity
Tongue
RHOMBENCEPHALIC SUPERVENTRICLE
(FUTURE FOURTH VENTRICLE)
CEREBELLUM
Mandibular process
Oral cavity
Tongue
Basal occipital
UPPER MEDULLA
RHOMBENCEPHALON
Medullary velum
Epi-glottis
Rhombencephalic choroid plexus
Larynx
Pharynx
Atlas
Axis
Vertebral column
LOWER MEDULLA
Dorsal root ganglion
Future squamous occipital bone
FONT KEY:
VENTRICULAR DIVISIONS - CAPITALS
Major brain structure - Times **Bold CAPITALS**
All other structures - Times Roman or **Bold**

PLATE 2C

CR 21 mm, GW 8, C6202
Sagittal, Slide 26, Section 2
NEUROEPITHELIAL AND PARENCHYMAL BRAIN STRUCTURES

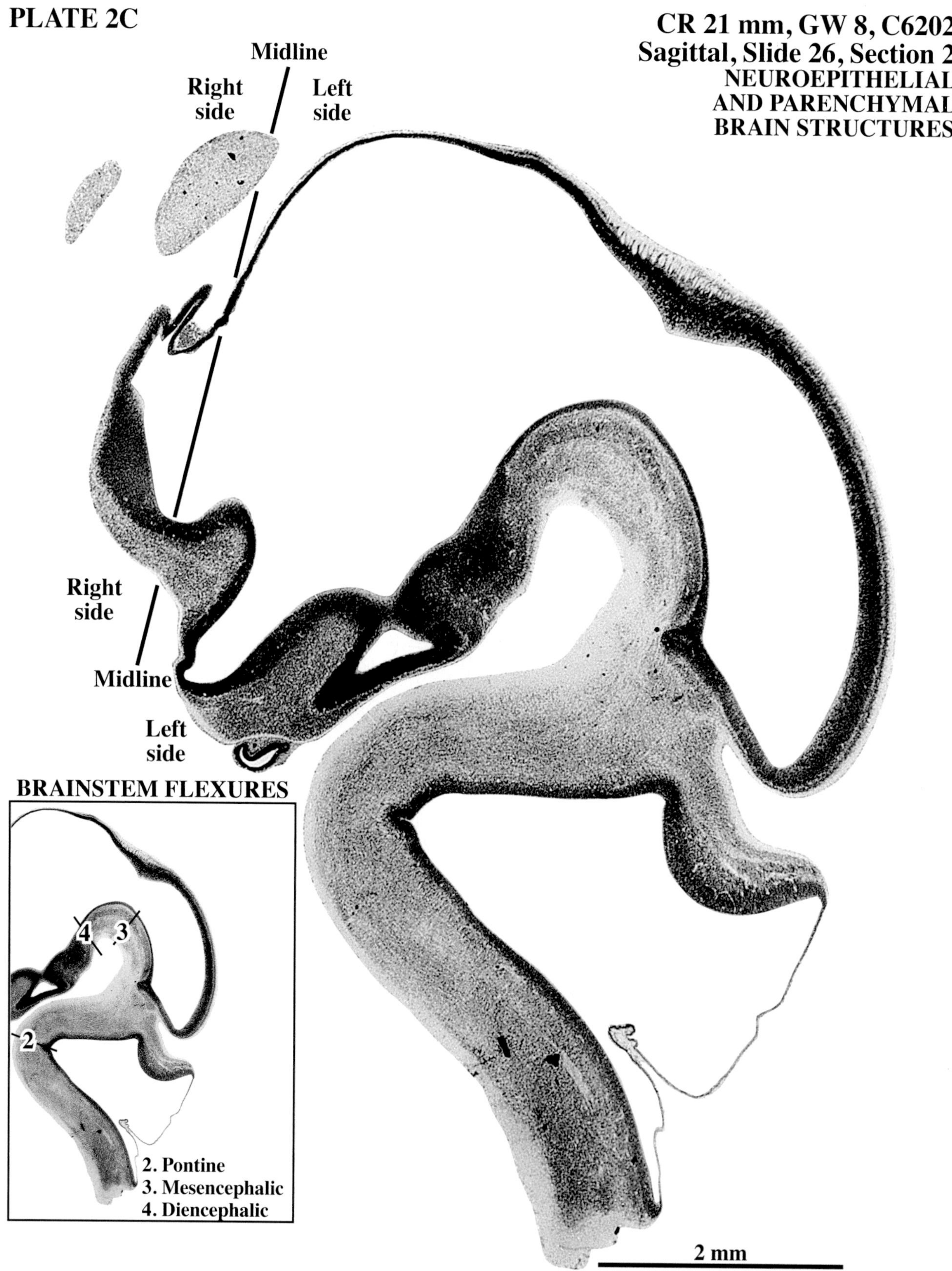
Midline
Right side
Left side
Right side
Midline
Left side
BRAINSTEM FLEXURES
4
3
2
2. Pontine
3. Mesencephalic
4. Diencephalic
2 mm

The skull, major brain structures, and ventricular divisions are labeled in Parts A and B of this plate on the preceding pages.

See Plate 4 for details of the anterior and middle hypothalamus, Plate 12 for details of the midbrain tegmentum in a nearby section.

PLATE 2D

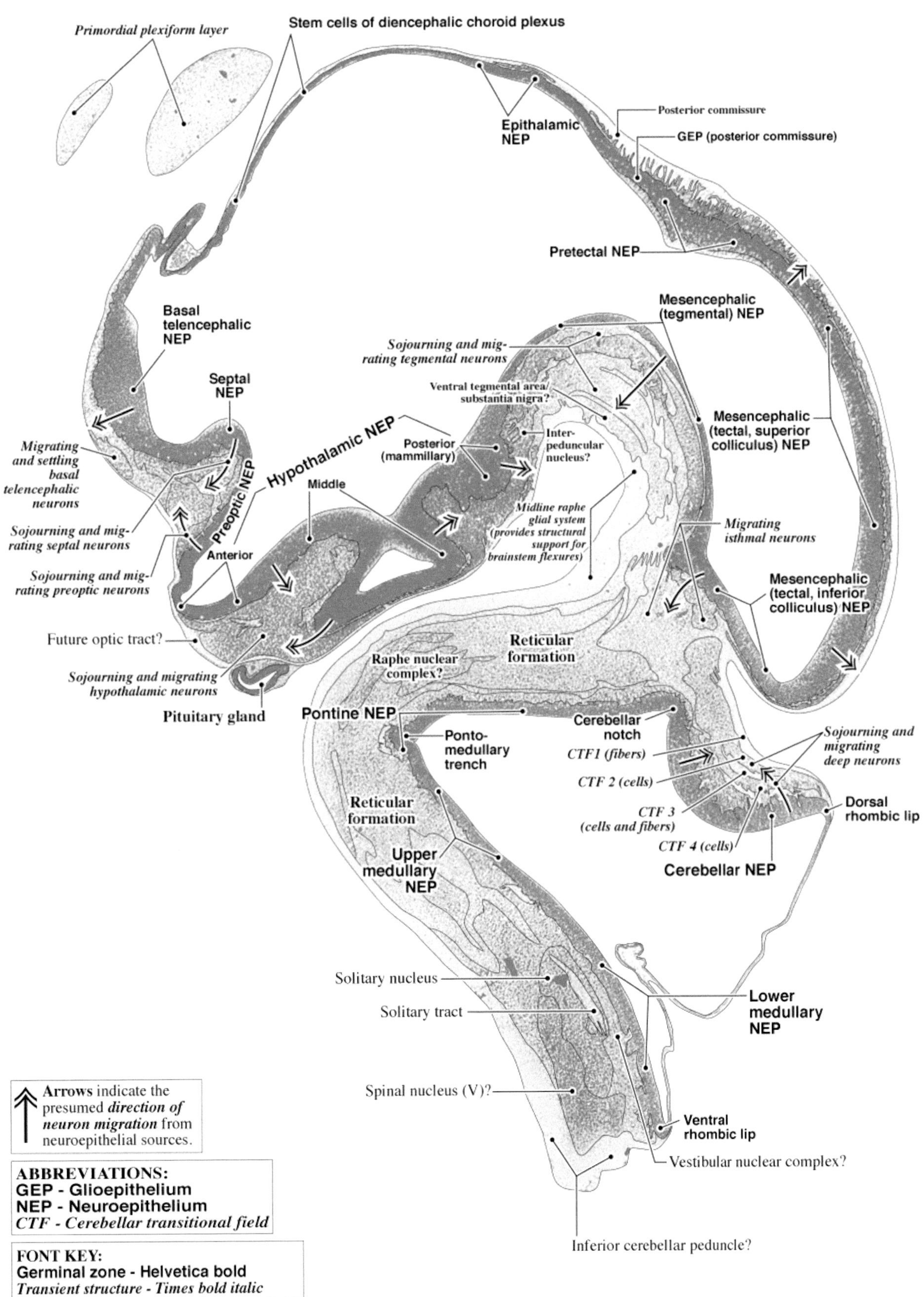
Primordial plexiform layer
Stem cells of diencephalic choroid plexus
Epithalamic NEP
Posterior commissure
GEP (posterior commissure)
Pretectal NEP
Mesencephalic (tegmental) NEP
Basal telencephalic NEP
Septal NEP
Sojourning and migrating tegmental neurons
Ventral tegmental area/ substantia nigra?
Mesencephalic (tectal, superior colliculus) NEP
Migrating and settling basal telencephalic neurons
Hypothalamic NEP
Posterior (mammillary)
Inter-peduncular nucleus?
Preoptic NEP
Middle
Midline raphe glial system (provides structural support for brainstem flexures)
Migrating isthmal neurons
Sojourning and migrating septal neurons
Anterior
Sojourning and migrating preoptic neurons
Mesencephalic (tectal, inferior colliculus) NEP
Future optic tract?
Reticular formation
Raphe nuclear complex?
Sojourning and migrating hypothalamic neurons
Pituitary gland
Pontine NEP
Cerebellar notch
Ponto-medullary trench
CTF1 (fibers)
Sojourning and migrating deep neurons
CTF 2 (cells)
Dorsal rhombic lip
Reticular formation
CTF 3 (cells and fibers)
CTF 4 (cells)
Upper medullary NEP
Cerebellar NEP
Solitary nucleus
Lower medullary NEP
Solitary tract
Spinal nucleus (V)?
Ventral rhombic lip
Vestibular nuclear complex?
Inferior cerebellar peduncle?

Arrows indicate the presumed ***direction of neuron migration*** from neuroepithelial sources.

ABBREVIATIONS:
GEP - Glioepithelium
NEP - Neuroepithelium
CTF - Cerebellar transitional field

FONT KEY:
Germinal zone - Helvetica bold
Transient structure - Times bold italic
Permanent structure - Times Roman or **Bold**

PLATE 3A

CR 21 mm, GW 8, C6202
Sagittal, Slide 24, Section 2
Left side of brain
SKULL, MAJOR BRAIN STRUCTURES, AND VENTRICULAR DIVISIONS

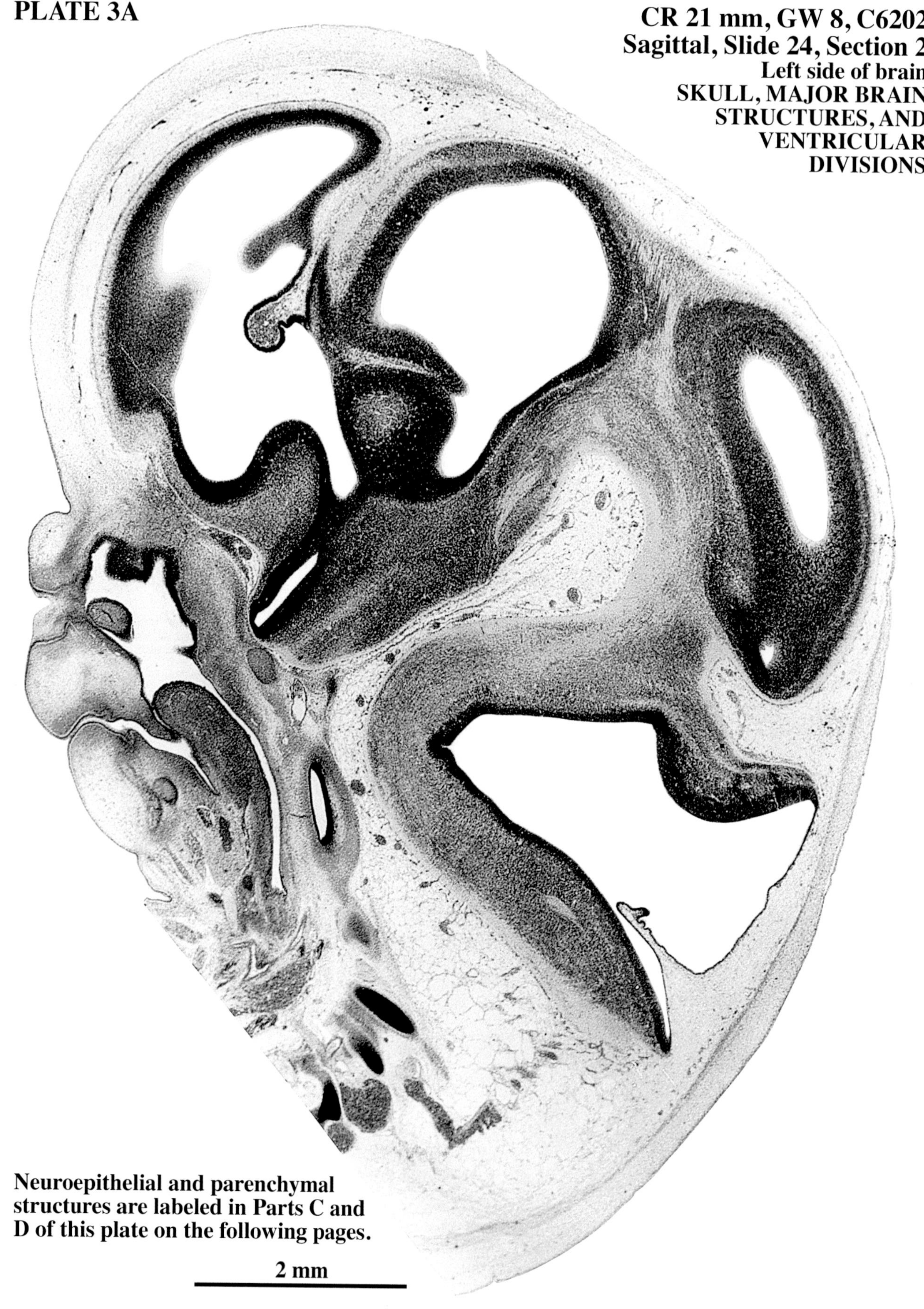

Neuroepithelial and parenchymal structures are labeled in Parts C and D of this plate on the following pages.

2 mm

PLATE 3B

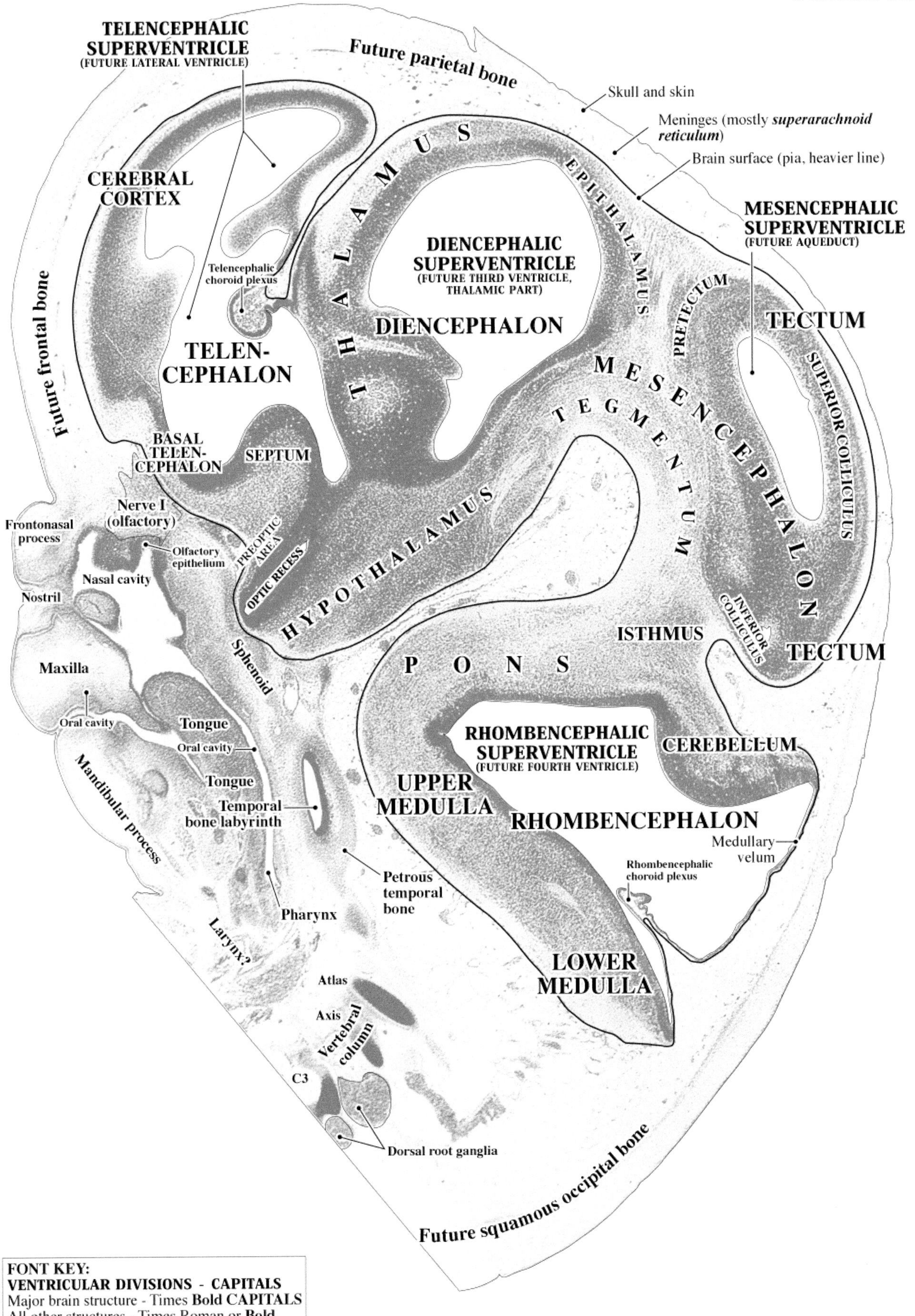
TELENCEPHALIC SUPERVENTRICLE
(FUTURE LATERAL VENTRICLE)
Future parietal bone
Skull and skin
Meninges (mostly superarachnoid reticulum)
Brain surface (pia, heavier line)
CEREBRAL CORTEX
THALAMUS
EPITHALAMUS
DIENCEPHALIC SUPERVENTRICLE
(FUTURE THIRD VENTRICLE, THALAMIC PART)
MESENCEPHALIC SUPERVENTRICLE
(FUTURE AQUEDUCT)
Telencephalic choroid plexus
DIENCEPHALON
PRETECTUM
TECTUM
Future frontal bone
TELEN-CEPHALON
MESENCEPHALON
SUPERIOR COLLICULUS
TEGMENTUM
BASAL TELEN-CEPHALON
SEPTUM
HYPOTHALAMUS
Nerve I (olfactory)
Frontonasal process
Olfactory epithelium
PREOPTIC AREA
OPTIC RECESS
Nasal cavity
Nostril
INFERIOR COLLICULUS
ISTHMUS
TECTUM
Maxilla
Sphenoid
PONS
Oral cavity
Tongue
Oral cavity
RHOMBENCEPHALIC SUPERVENTRICLE
(FUTURE FOURTH VENTRICLE)
CEREBELLUM
Tongue
Mandibular process
UPPER MEDULLA
Temporal bone labyrinth
RHOMBENCEPHALON
Medullary velum
Rhombencephalic choroid plexus
Petrous temporal bone
Pharynx
Larynx?
LOWER MEDULLA
Atlas
Axis
Vertebral column
C3
Dorsal root ganglia
Future squamous occipital bone

FONT KEY:
VENTRICULAR DIVISIONS - CAPITALS
Major brain structure - Times **Bold CAPITALS**
All other structures - Times Roman or **Bold**

PLATE 3C

CR 21 mm, GW 8, C6202
Sagittal, Slide 24, Section 2
Left side of brain
NEUROEPITHELIAL AND PARENCHYMAL BRAIN STRUCTURES

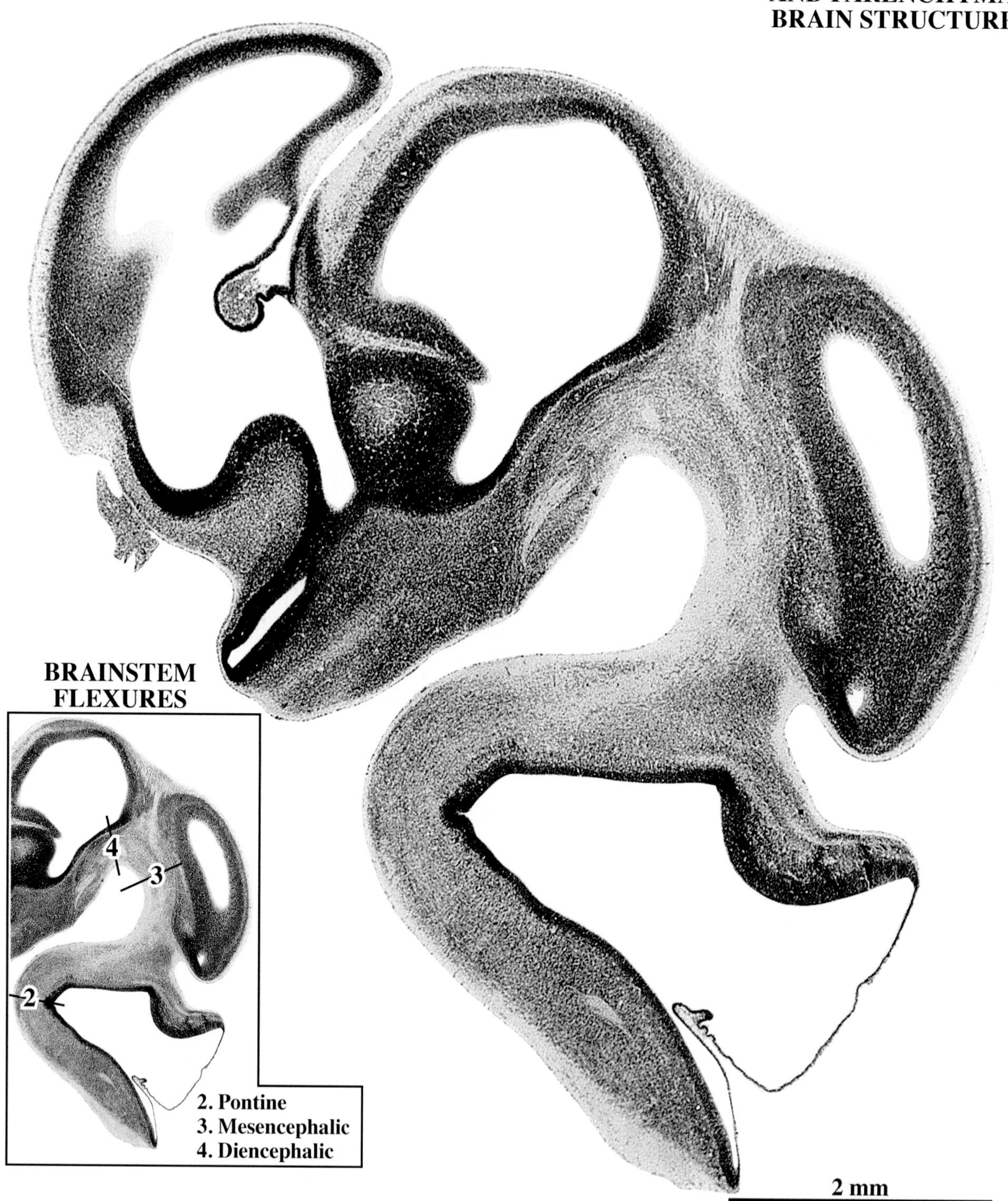
BRAINSTEM
FLEXURES
4
3
2
2. Pontine
3. Mesencephalic
4. Diencephalic
2 mm

The skull, major brain structures, and ventricular divisions are labeled in Parts A and B of this plate on the preceding pages.

PLATE 3D

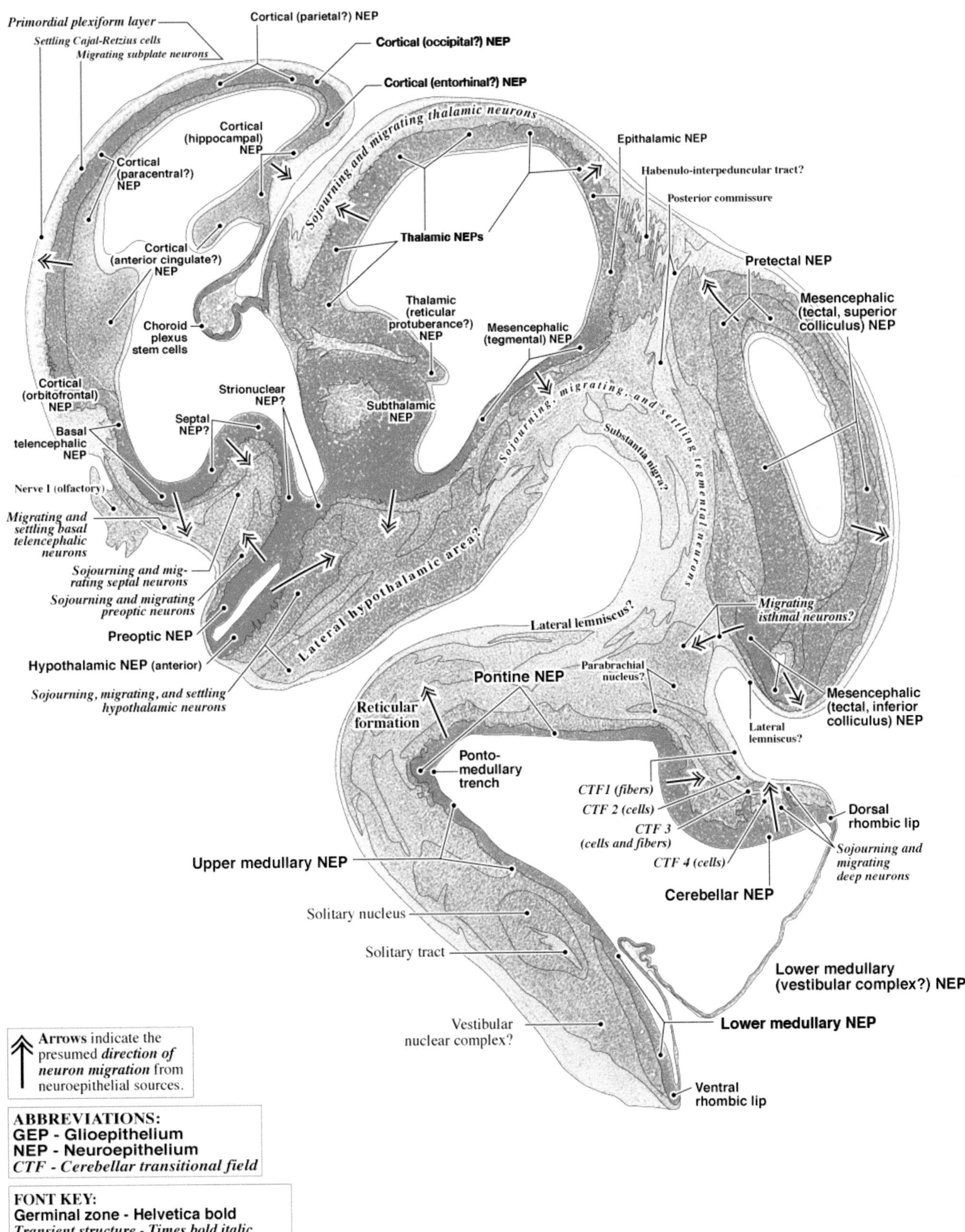
Primordial plexiform layer
Settling Cajal-Retzius cells
Migrating subplate neurons
Cortical (parietal?) NEP
Cortical (occipital?) NEP
Cortical (entorhinal?) NEP
Cortical (hippocampal) NEP
Cortical (paracentral?) NEP
Cortical (anterior cingulate?) NEP
Choroid plexus stem cells
Cortical (orbitofrontal) NEP
Basal telencephalic NEP
Nerve I (olfactory)
Migrating and settling basal telencephalic neurons
Sojourning and migrating septal neurons
Sojourning and migrating preoptic neurons
Preoptic NEP
Hypothalamic NEP (anterior)
Sojourning, migrating, and settling hypothalamic neurons
Septal NEP?
Strionuclear NEP?
Subthalamic NEP
Sojourning and migrating thalamic neurons
Thalamic NEPs
Thalamic (reticular protuberance?) NEP
Mesencephalic (tegmental) NEP
Lateral hypothalamic area?
Epithalamic NEP
Habenulo-interpeduncular tract?
Posterior commissure
Pretectal NEP
Mesencephalic (tectal, superior colliculus) NEP
Sojourning, migrating, and settling tegmental neurons
Substantia nigra?
Lateral lemniscus?
Migrating isthmal neurons?
Parabrachial nucleus?
Pontine NEP
Mesencephalic (tectal, inferior colliculus) NEP
Lateral lemniscus?
Reticular formation
Ponto-medullary trench
CTF1 (fibers)
CTF 2 (cells)
CTF 3 (cells and fibers)
CTF 4 (cells)
Dorsal rhombic lip
Sojourning and migrating deep neurons
Cerebellar NEP
Upper medullary NEP
Solitary nucleus
Solitary tract
Lower medullary (vestibular complex?) NEP
Lower medullary NEP
Vestibular nuclear complex?
Ventral rhombic lip

Arrows indicate the presumed ***direction of neuron migration*** from neuroepithelial sources.

ABBREVIATIONS:
GEP - Glioepithelium
NEP - Neuroepithelium
CTF - Cerebellar transitional field

FONT KEY:
Germinal zone - Helvetica bold
Transient structure - Times bold italic
Permanent structure - Times Roman or **Bold**

PLATE 4A

CR 21 mm, GW 8, C6202
Sagittal, Slide 22, Section 2
Left side of brain
SKULL, MAJOR BRAIN STRUCTURES, AND VENTRICULAR DIVISIONS

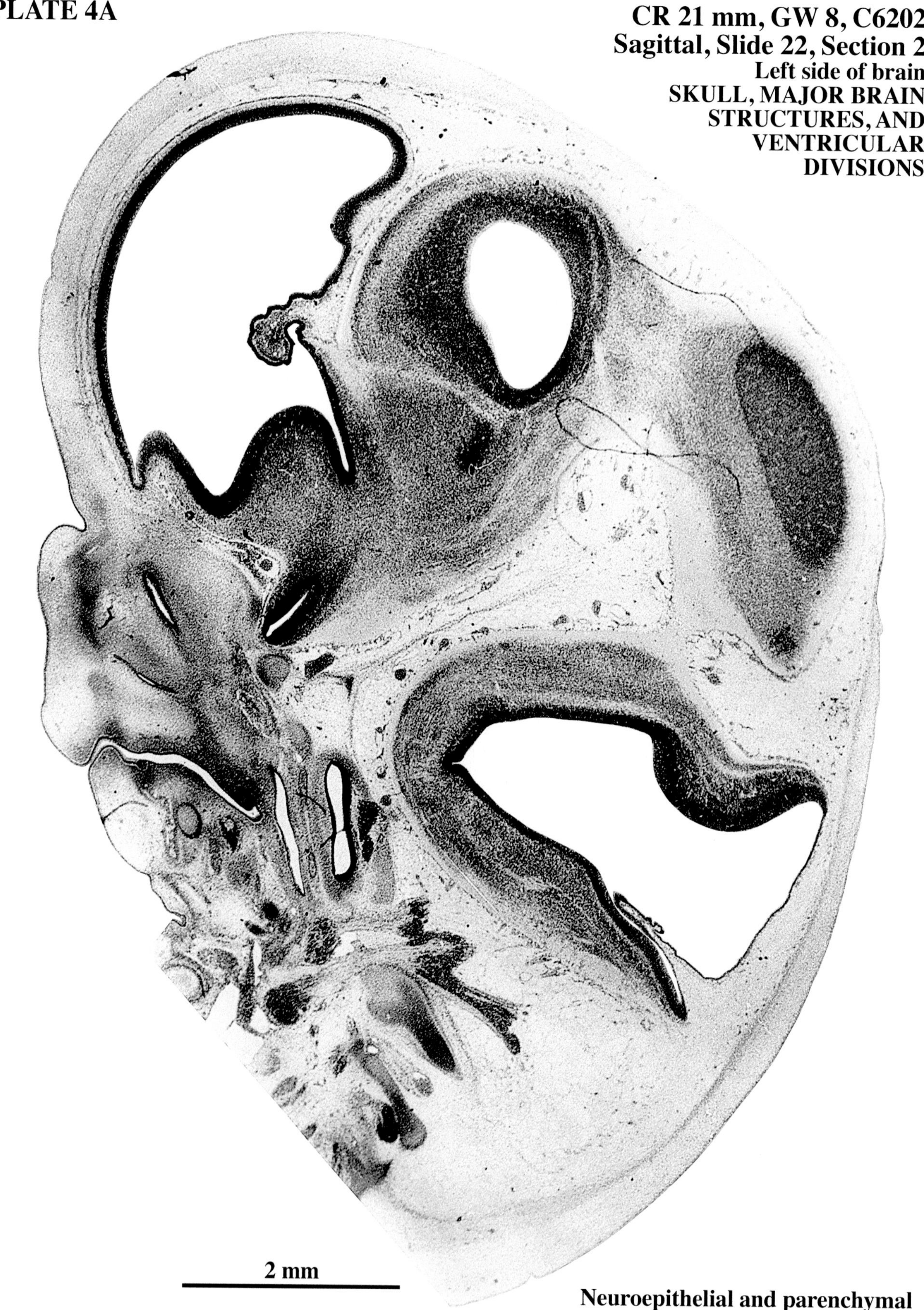
2 mm

Neuroepithelial and parenchymal structures are labeled in Parts C and D of this plate on the following pages.

PLATE 4B

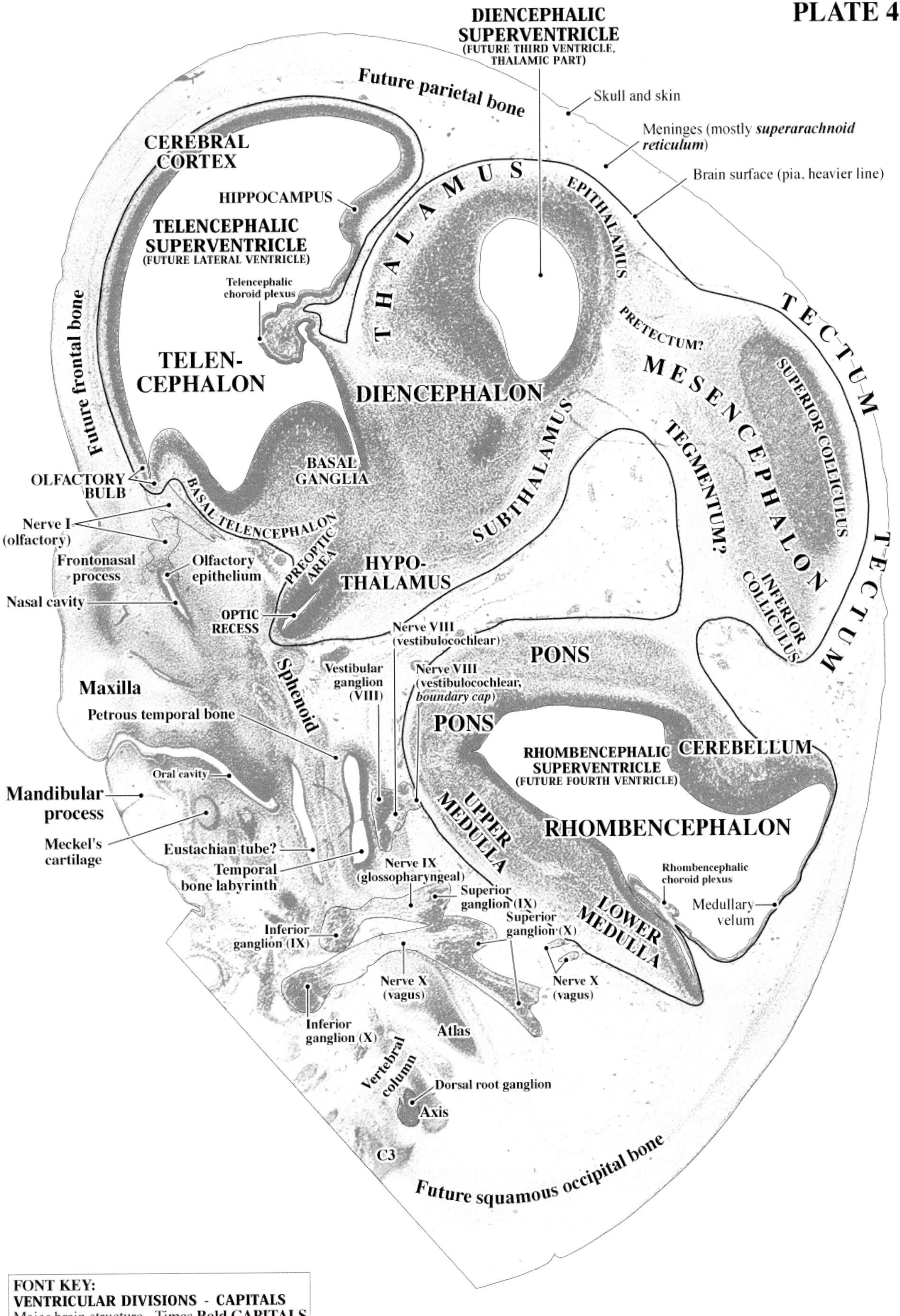
DIENCEPHALIC SUPERVENTRICLE
(FUTURE THIRD VENTRICLE, THALAMIC PART)
Future parietal bone
Skull and skin
Meninges (mostly *superarachnoid reticulum*)
Brain surface (pia, heavier line)
CEREBRAL CORTEX
HIPPOCAMPUS
TELENCEPHALIC SUPERVENTRICLE
(FUTURE LATERAL VENTRICLE)
Telencephalic choroid plexus
THALAMUS
EPITHALAMUS
TECTUM
PRETECTUM?
Future frontal bone
TELEN-CEPHALON
MESENCEPHALON
SUPERIOR COLLICULUS
DIENCEPHALON
TEGMENTUM?
SUBTHALAMUS
BASAL GANGLIA
OLFACTORY BULB
BASAL TELENCEPHALON
Nerve I (olfactory)
PREOPTIC AREA
Frontonasal process
Olfactory epithelium
HYPO-THALAMUS
INFERIOR COLLICULUS
Nasal cavity
OPTIC RECESS
Nerve VIII (vestibulocochlear)
PONS
Vestibular ganglion (VIII)
Nerve VIII (vestibulocochlear, *boundary cap*)
Maxilla
Sphenoid
Petrous temporal bone
PONS
RHOMBENCEPHALIC SUPERVENTRICLE
(FUTURE FOURTH VENTRICLE)
CEREBELLUM
Oral cavity
Mandibular process
UPPER MEDULLA
RHOMBENCEPHALON
Meckel's cartilage
Eustachian tube?
Temporal bone labyrinth
Nerve IX (glossopharyngeal)
Rhombencephalic choroid plexus
Superior ganglion (IX)
Medullary velum
Superior ganglion (X)
LOWER MEDULLA
Inferior ganglion (IX)
Nerve X (vagus)
Nerve X (vagus)
Inferior ganglion (X)
Atlas
Vertebral column
Dorsal root ganglion
Axis
C3
Future squamous occipital bone
FONT KEY:
VENTRICULAR DIVISIONS - CAPITALS
Major brain structure - Times Bold CAPITALS
All other structures - Times Roman or Bold

PLATE 4C

CR 21 mm, GW 8, C6202
Sagittal, Slide 22, Section 2
Left side of brain
NEUROEPITHELIAL AND PARENCHYMAL BRAIN STRUCTURES, PERIPHERAL GANGLIA

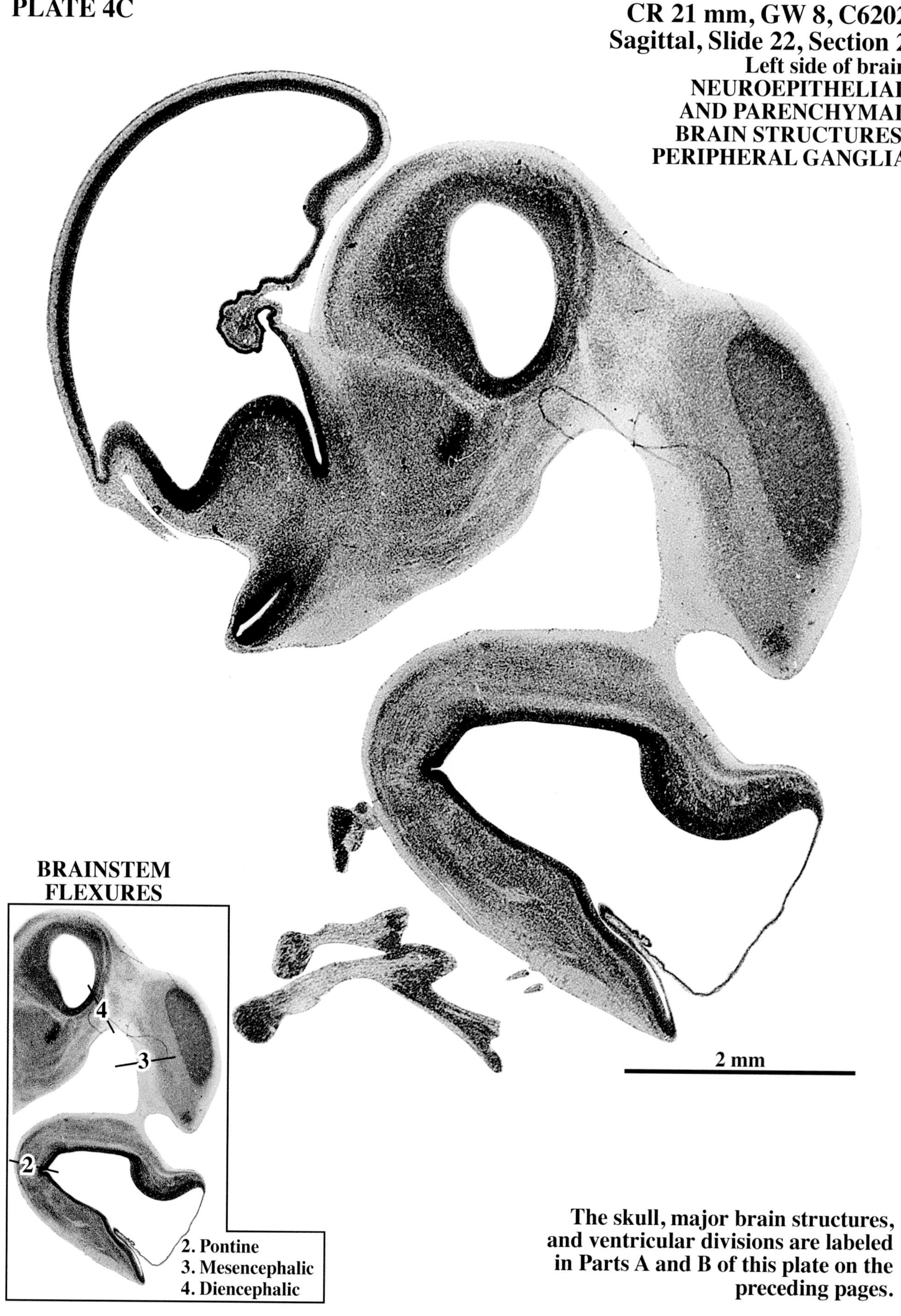

2 mm
BRAINSTEM FLEXURES
4
3
2
2. Pontine
3. Mesencephalic
4. Diencephalic

The skull, major brain structures, and ventricular divisions are labeled in Parts A and B of this plate on the preceding pages.

PLATE 4D

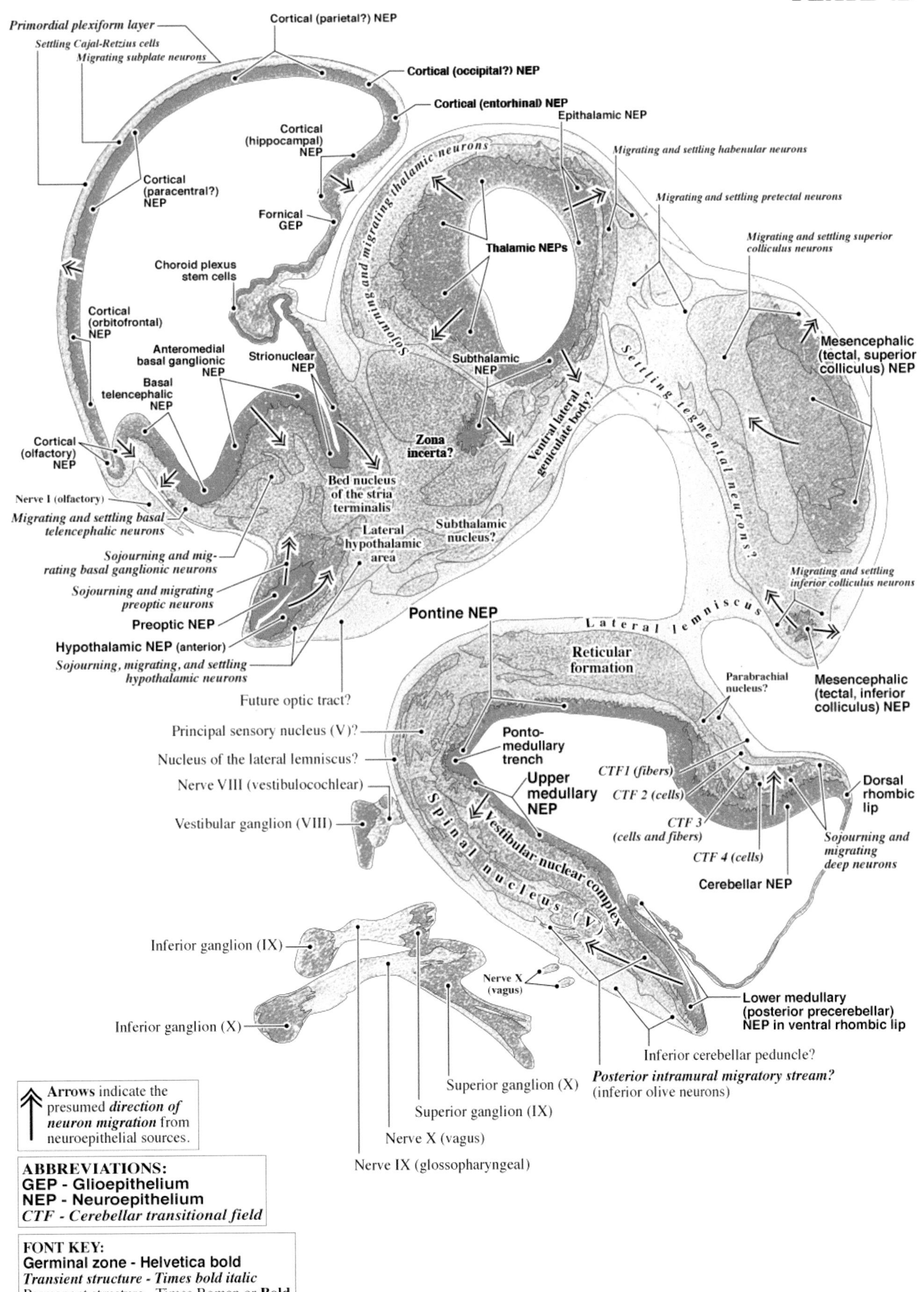

Primordial plexiform layer
Settling Cajal-Retzius cells
Migrating subplate neurons
Cortical (parietal?) NEP
Cortical (occipital?) NEP
Cortical (entorhinal) NEP
Epithalamic NEP
Migrating and settling habenular neurons
Migrating and settling pretectal neurons
Migrating and settling superior colliculus neurons
Cortical (hippocampal) NEP
Cortical (paracentral?) NEP
Fornical GEP
Sojourning and migrating thalamic neurons
Thalamic NEPs
Choroid plexus stem cells
Cortical (orbitofrontal) NEP
Anteromedial basal ganglionic NEP
Strionuclear NEP
Subthalamic NEP
Settling tegmental neurons?
Mesencephalic (tectal, superior colliculus) NEP
Basal telencephalic NEP
Cortical (olfactory) NEP
Zona incerta?
Ventral lateral geniculate body?
Bed nucleus of the stria terminalis
Nerve I (olfactory)
Migrating and settling basal telencephalic neurons
Lateral hypothalamic area
Subthalamic nucleus?
Sojourning and migrating basal ganglionic neurons
Sojourning and migrating preoptic neurons
Migrating and settling inferior colliculus neurons
Preoptic NEP
Pontine NEP
Lateral lemniscus
Hypothalamic NEP (anterior)
Sojourning, migrating, and settling hypothalamic neurons
Reticular formation
Parabrachial nucleus?
Mesencephalic (tectal, inferior colliculus) NEP
Future optic tract?
Principal sensory nucleus (V)?
Ponto-medullary trench
Nucleus of the lateral lemniscus?
CTF1 (fibers)
Dorsal rhombic lip
Nerve VIII (vestibulocochlear)
Upper medullary NEP
CTF 2 (cells)
Vestibular ganglion (VIII)
CTF 3 (cells and fibers)
Sojourning and migrating deep neurons
CTF 4 (cells)
Spinal nucleus (V)
Vestibular nuclear complex
Cerebellar NEP
Inferior ganglion (IX)
Nerve X (vagus)
Lower medullary (posterior precerebellar) NEP in ventral rhombic lip
Inferior ganglion (X)
Inferior cerebellar peduncle?
Posterior intramural migratory stream? (inferior olive neurons)
Superior ganglion (X)
Superior ganglion (IX)
Nerve X (vagus)
Nerve IX (glossopharyngeal)

Arrows indicate the presumed ***direction of neuron migration*** from neuroepithelial sources.

ABBREVIATIONS:
GEP - Glioepithelium
NEP - Neuroepithelium
CTF - Cerebellar transitional field

FONT KEY:
Germinal zone - Helvetica bold
Transient structure - Times bold italic
Permanent structure - Times Roman or **Bold**

PLATE 5A

CR 21 mm, GW 8, C6202
Sagittal, Slide 21, Section 2
Left side of brain
SKULL, MAJOR BRAIN STRUCTURES, AND VENTRICULAR DIVISIONS

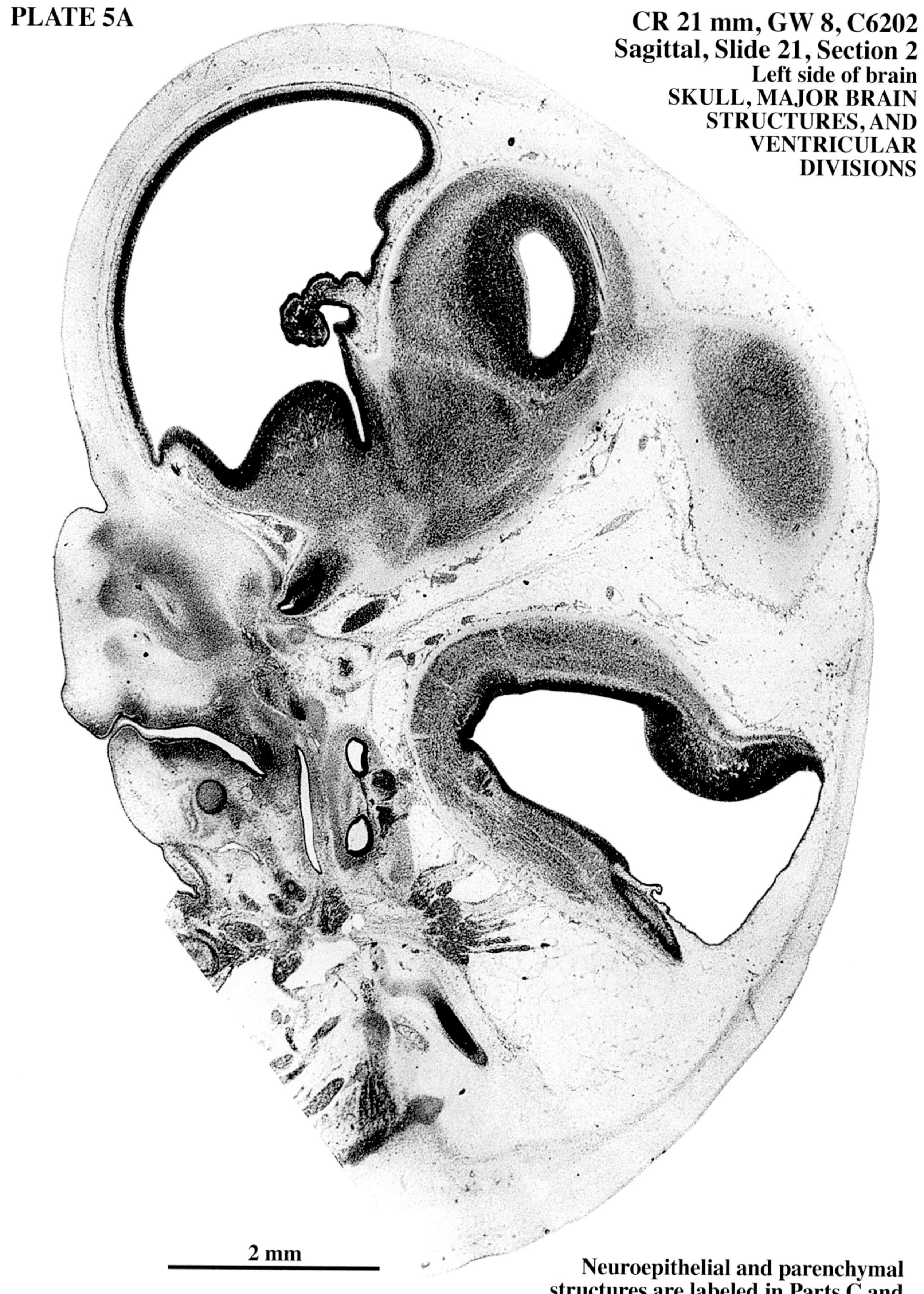
2 mm

Neuroepithelial and parenchymal structures are labeled in Parts C and D of this plate on the following pages.

PLATE 5B

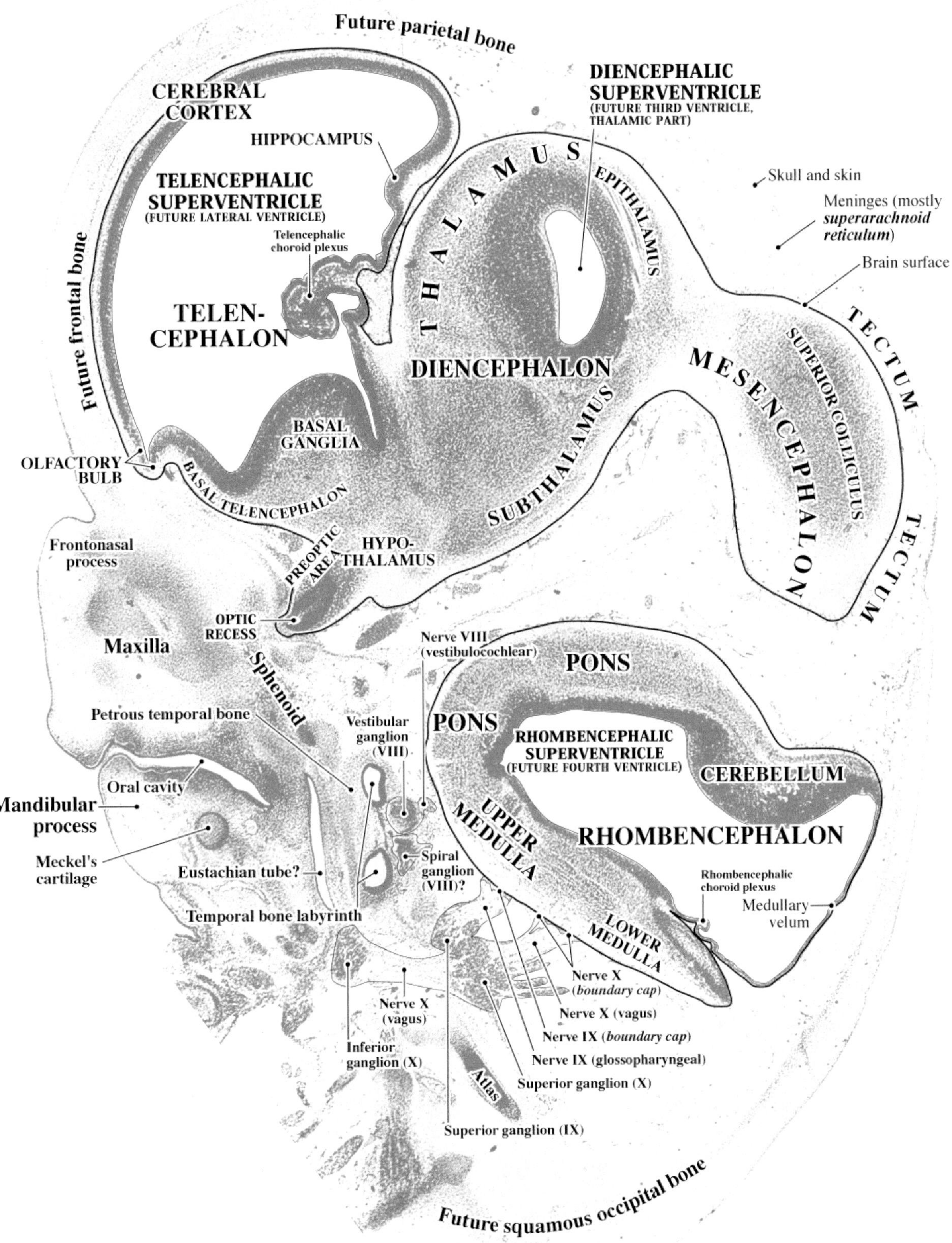
Future parietal bone
CEREBRAL CORTEX
HIPPOCAMPUS
TELENCEPHALIC SUPERVENTRICLE
(FUTURE LATERAL VENTRICLE)
Telencephalic choroid plexus
TELEN-CEPHALON
Future frontal bone
BASAL GANGLIA
OLFACTORY BULB
BASAL TELENCEPHALON
DIENCEPHALIC SUPERVENTRICLE
(FUTURE THIRD VENTRICLE, THALAMIC PART)
THALAMUS
EPITHALAMUS
DIENCEPHALON
SUBTHALAMUS
PREOPTIC AREA
HYPO-THALAMUS
OPTIC RECESS
Skull and skin
Meninges (mostly *superarachnoid reticulum*)
Brain surface
TECTUM
SUPERIOR COLLICULUS
MESENCEPHALON
TECTUM
Frontonasal process
Maxilla
Sphenoid
Petrous temporal bone
Oral cavity
Mandibular process
Meckel's cartilage
Eustachian tube?
Temporal bone labyrinth
Vestibular ganglion (VIII)
Nerve VIII (vestibulocochlear)
Spiral ganglion (VIII)?
PONS
PONS
RHOMBENCEPHALIC SUPERVENTRICLE
(FUTURE FOURTH VENTRICLE)
CEREBELLUM
UPPER MEDULLA
RHOMBENCEPHALON
Rhombencephalic choroid plexus
Medullary velum
LOWER MEDULLA
Nerve X (*boundary cap*)
Nerve X (vagus)
Nerve IX (*boundary cap*)
Nerve IX (glossopharyngeal)
Superior ganglion (X)
Superior ganglion (IX)
Nerve X (vagus)
Inferior ganglion (X)
Atlas
Future squamous occipital bone

FONT KEY:
VENTRICULAR DIVISIONS - CAPITALS
Major brain structure - Times **Bold CAPITALS**
All other structures - Times Roman or **Bold**

PLATE 5C

CR 21 mm, GW 8, C6202
Sagittal, Slide 21, Section 2
Left side of brain
NEUROEPITHELIAL AND PARENCHYMAL BRAIN STRUCTURES, PERIPHERAL GANGLIA

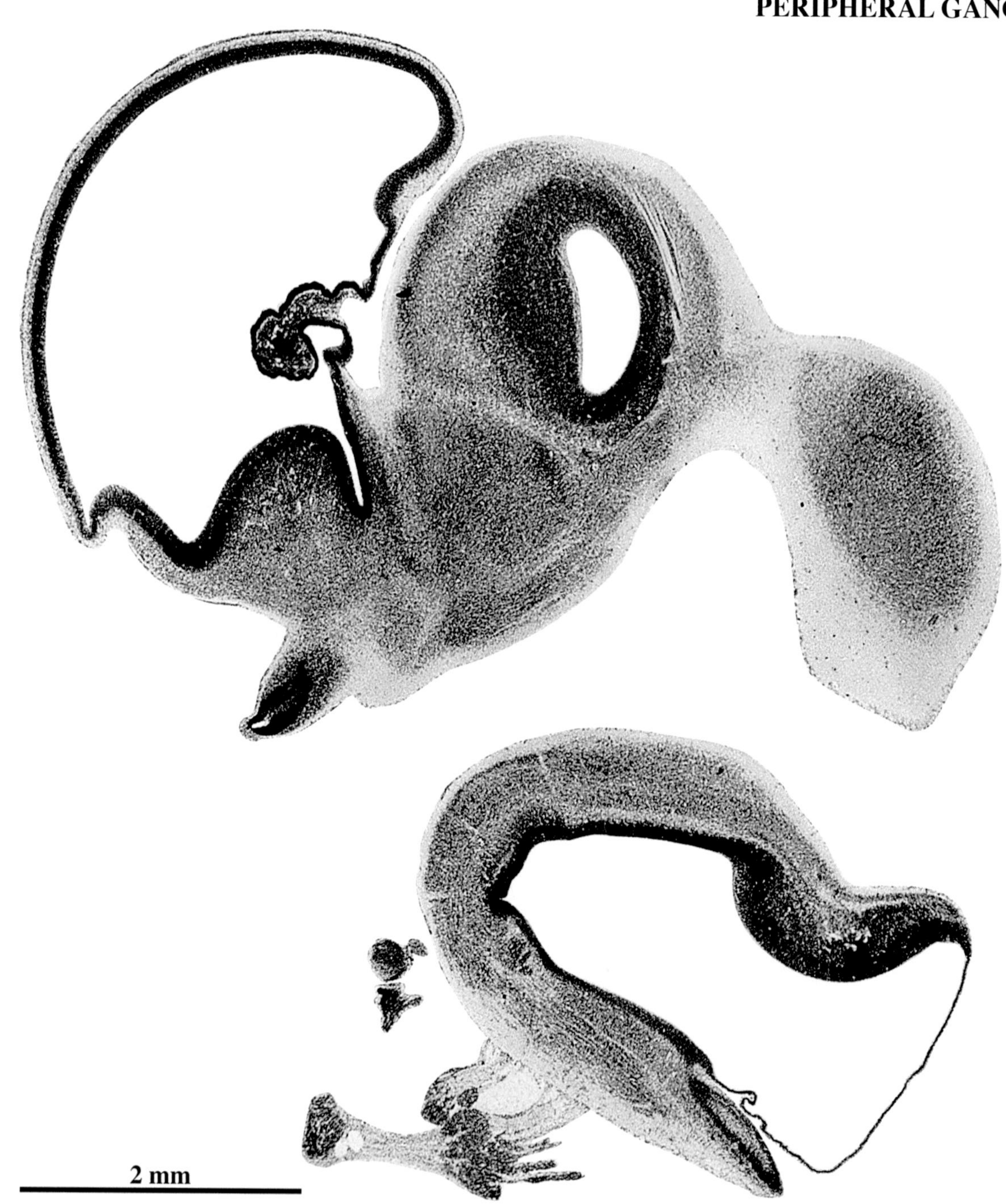
2 mm

The skull, major brain structures, and ventricular divisions are labeled in Parts A and B of this plate on the preceding pages.

PLATE 5D

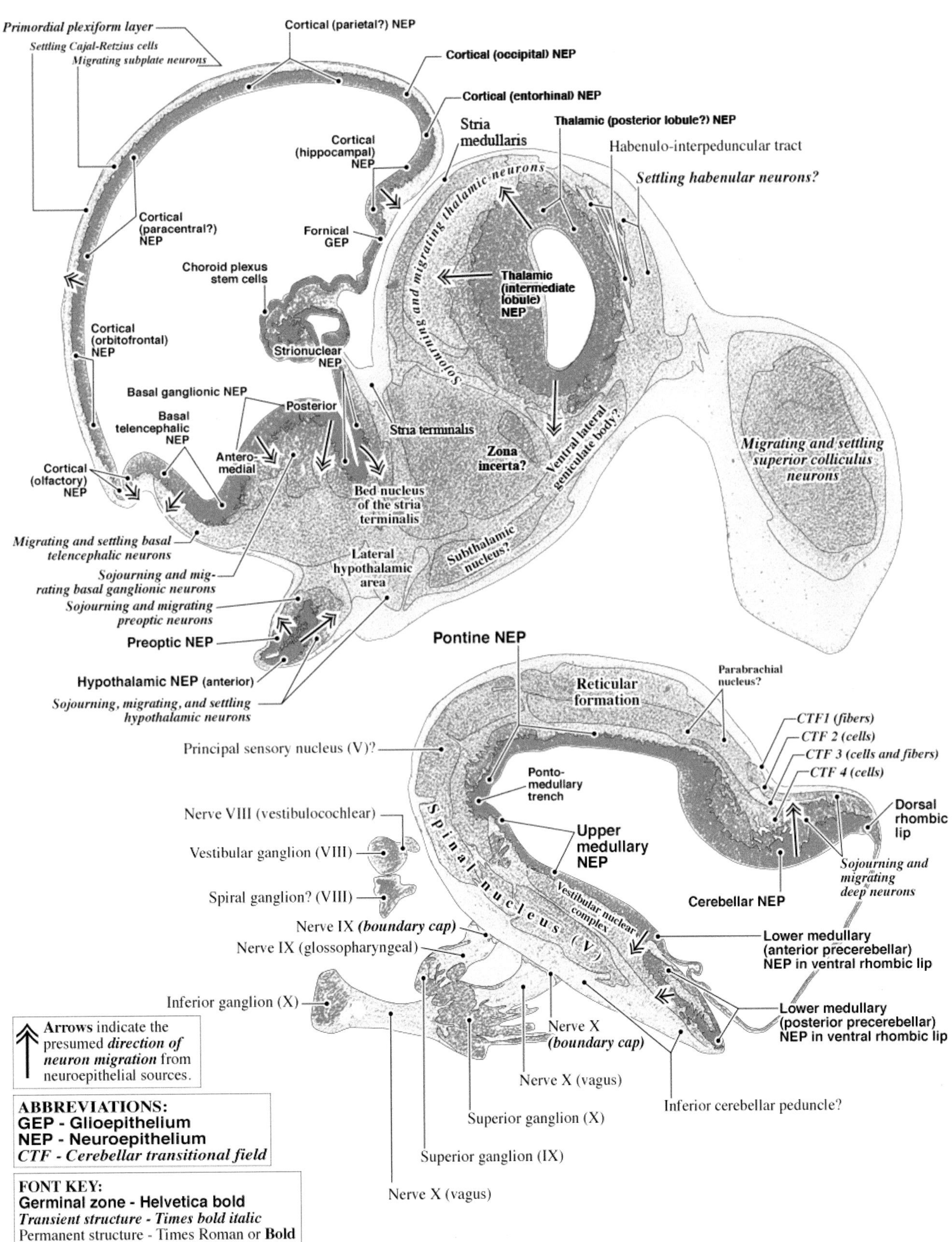
Primordial plexiform layer
Settling Cajal-Retzius cells
Migrating subplate neurons
Cortical (parietal?) NEP
Cortical (occipital) NEP
Cortical (entorhinal) NEP
Stria medullaris
Thalamic (posterior lobule?) NEP
Habenulo-interpeduncular tract
Settling habenular neurons?
Cortical (hippocampal) NEP
Cortical (paracentral?) NEP
Fornical GEP
Choroid plexus stem cells
Sojourning and migrating thalamic neurons
Thalamic (intermediate lobule) NEP
Cortical (orbitofrontal) NEP
Strionuclear NEP
Basal ganglionic NEP
Posterior
Basal telencephalic NEP
Stria terminalis
Antero-medial
Zona incerta?
Ventral lateral geniculate body?
Migrating and settling superior colliculus neurons
Cortical (olfactory) NEP
Bed nucleus of the stria terminalis
Migrating and settling basal telencephalic neurons
Lateral hypothalamic area
Subthalamic nucleus?
Sojourning and migrating basal ganglionic neurons
Sojourning and migrating preoptic neurons
Preoptic NEP
Pontine NEP
Hypothalamic NEP (anterior)
Sojourning, migrating, and settling hypothalamic neurons
Reticular formation
Parabrachial nucleus?
CTF1 (fibers)
CTF 2 (cells)
CTF 3 (cells and fibers)
CTF 4 (cells)
Principal sensory nucleus (V)?
Ponto-medullary trench
Dorsal rhombic lip
Nerve VIII (vestibulocochlear)
Vestibular ganglion (VIII)
Upper medullary NEP
Sojourning and migrating deep neurons
Spinal nucleus (V)
Vestibular nuclear complex
Spiral ganglion? (VIII)
Cerebellar NEP
Nerve IX (boundary cap)
Nerve IX (glossopharyngeal)
Lower medullary (anterior precerebellar) NEP in ventral rhombic lip
Inferior ganglion (X)
Lower medullary (posterior precerebellar) NEP in ventral rhombic lip
Nerve X (boundary cap)
Nerve X (vagus)
Inferior cerebellar peduncle?
Superior ganglion (X)
Superior ganglion (IX)
Nerve X (vagus)

Arrows indicate the presumed ***direction of neuron migration*** from neuroepithelial sources.

ABBREVIATIONS:
GEP - Glioepithelium
NEP - Neuroepithelium
CTF - Cerebellar transitional field

FONT KEY:
Germinal zone - Helvetica bold
Transient structure - Times bold italic
Permanent structure - Times Roman or **Bold**

PLATE 6A

CR 21 mm, GW 8, C6202
Sagittal, Slide 18, Section 2
Left side of brain
SKULL, MAJOR BRAIN STRUCTURES, AND VENTRICULAR DIVISIONS

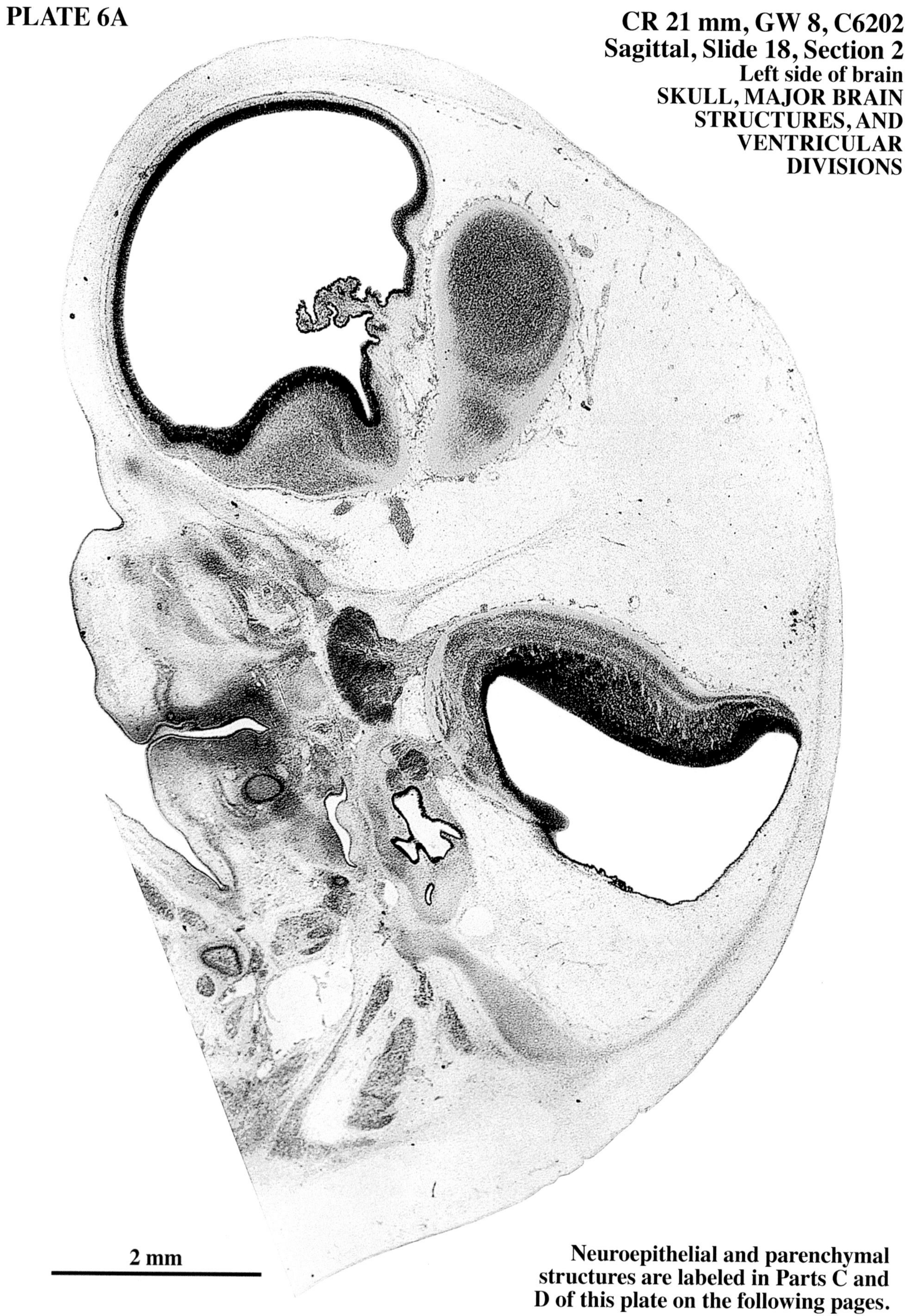
2 mm

Neuroepithelial and parenchymal structures are labeled in Parts C and D of this plate on the following pages.

PLATE 6B

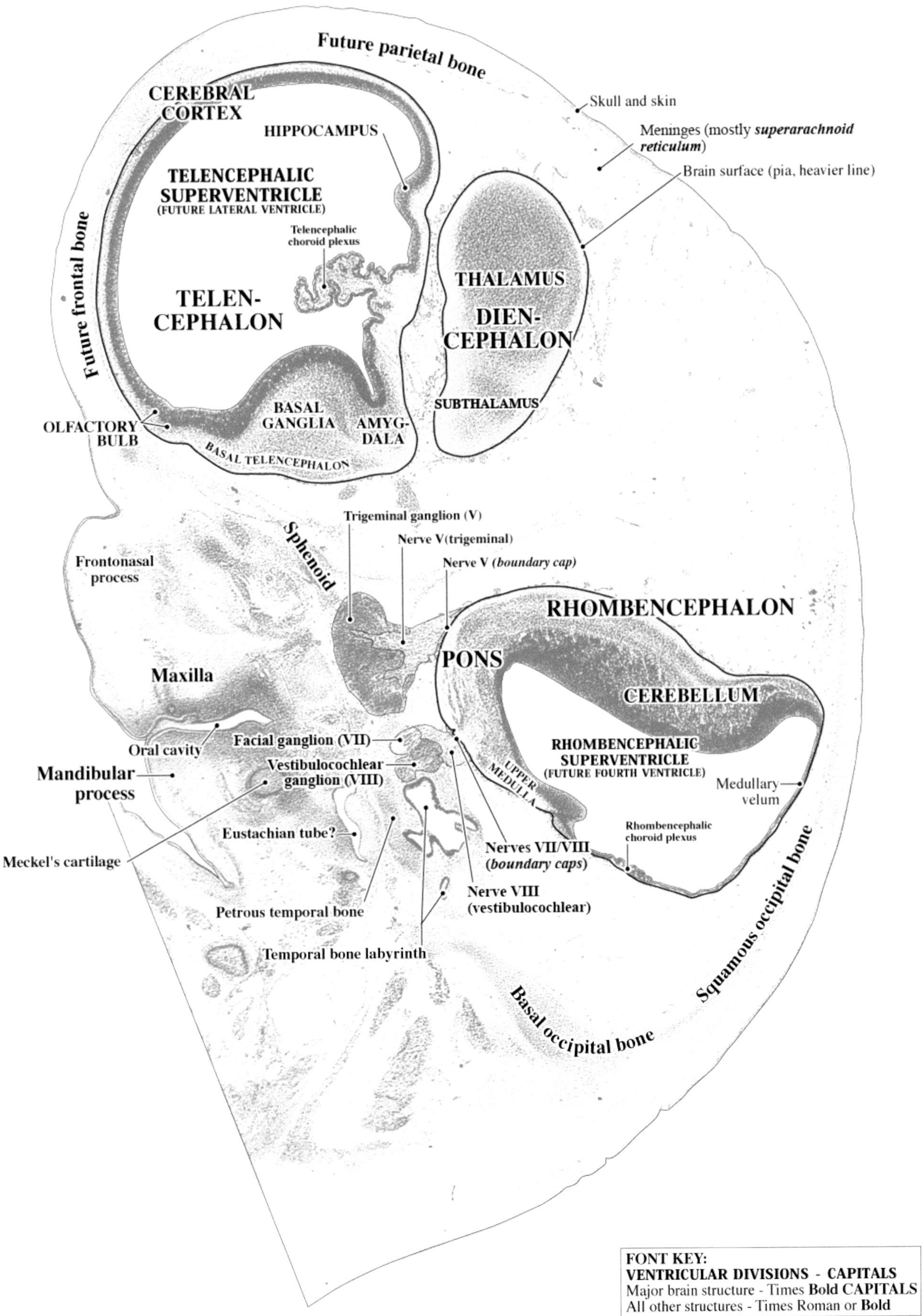
Future parietal bone
CEREBRAL CORTEX
HIPPOCAMPUS
Skull and skin
Meninges (mostly superarachnoid reticulum)
Brain surface (pia, heavier line)
TELENCEPHALIC SUPERVENTRICLE
(FUTURE LATERAL VENTRICLE)
Telencephalic choroid plexus
Future frontal bone
TELEN-CEPHALON
THALAMUS
DIEN-CEPHALON
SUBTHALAMUS
BASAL GANGLIA
AMYG-DALA
OLFACTORY BULB
BASAL TELENCEPHALON
Trigeminal ganglion (V)
Sphenoid
Nerve V(trigeminal)
Nerve V (boundary cap)
Frontonasal process
RHOMBENCEPHALON
PONS
Maxilla
CEREBELLUM
Facial ganglion (VII)
Oral cavity
RHOMBENCEPHALIC SUPERVENTRICLE
(FUTURE FOURTH VENTRICLE)
Mandibular process
Vestibulocochlear ganglion (VIII)
UPPER MEDULLA
Medullary velum
Eustachian tube?
Rhombencephalic choroid plexus
Meckel's cartilage
Nerves VII/VIII (boundary caps)
Nerve VIII (vestibulocochlear)
Petrous temporal bone
Squamous occipital bone
Temporal bone labyrinth
Basal occipital bone
FONT KEY:
VENTRICULAR DIVISIONS - CAPITALS
Major brain structure - Times Bold CAPITALS
All other structures - Times Roman or Bold

PLATE 6C

CR 21 mm, GW 8, C6202
Sagittal, Slide 18, Section 2
Left side of brain
NEUROEPITHELIAL AND PARENCHYMAL BRAIN STRUCTURES, PERIPHERAL GANGLIA

See details of the cerebral cortex in Plate 8.

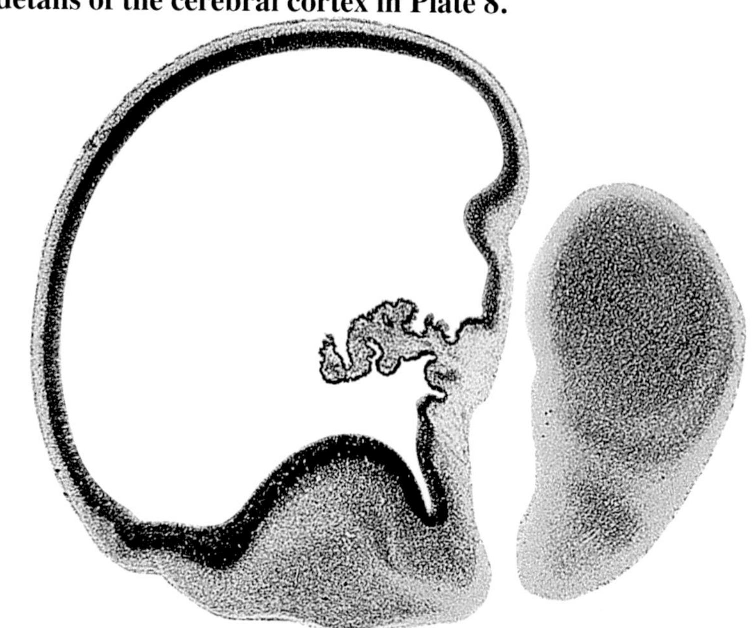

See details of the hippocampus, basal ganglia, and amygdala in Plate 10.

See details of the cerebellum in Plate 13.

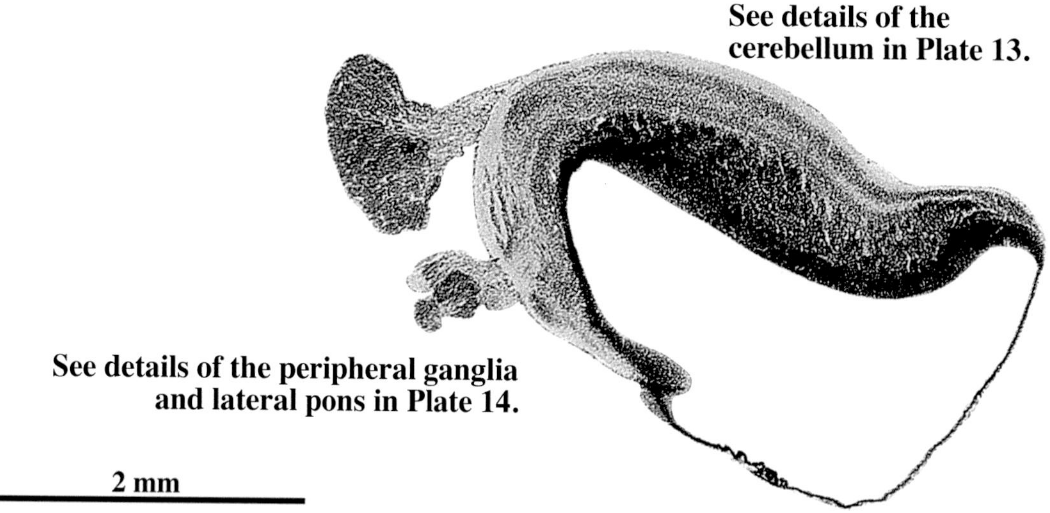

See details of the peripheral ganglia and lateral pons in Plate 14.

2 mm

The skull, major brain structures, and ventricular divisions are labeled in Parts A and B of this plate on the preceding pages.

PLATE 6D

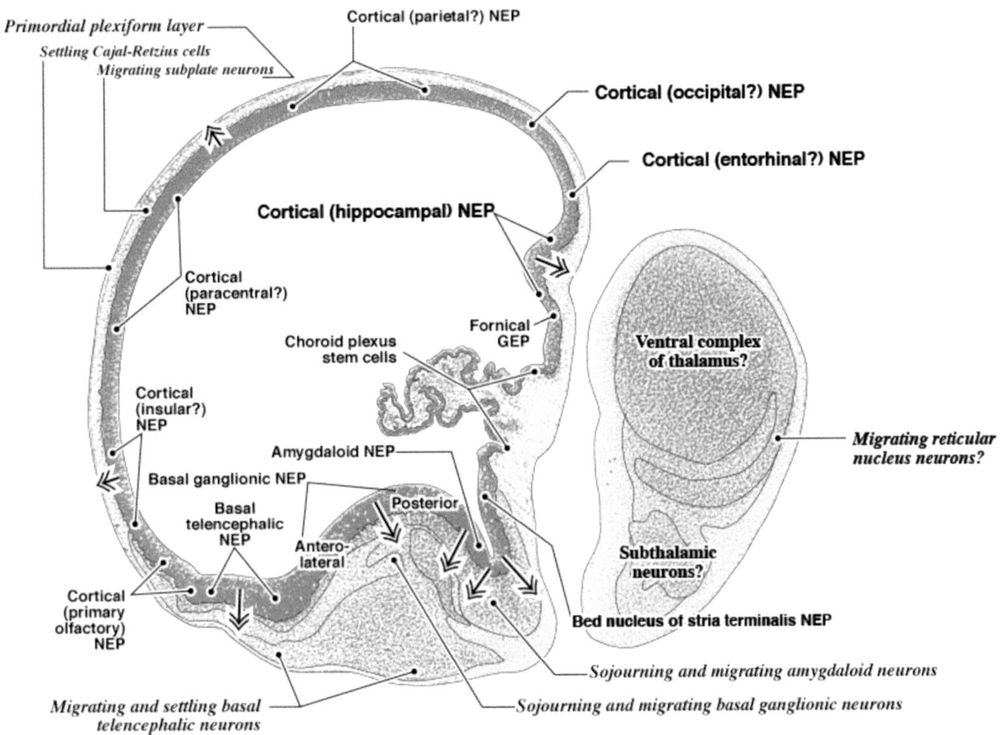

Primordial plexiform layer
Settling Cajal-Retzius cells
Migrating subplate neurons
Cortical (parietal?) NEP
Cortical (occipital?) NEP
Cortical (entorhinal?) NEP
Cortical (hippocampal) NEP
Cortical (paracentral?) NEP
Fornical GEP
Choroid plexus stem cells
Ventral complex of thalamus?
Cortical (insular?) NEP
Migrating reticular nucleus neurons?
Amygdaloid NEP
Basal ganglionic NEP
Posterior
Basal telencephalic NEP
Antero-lateral
Subthalamic neurons?
Cortical (primary olfactory) NEP
Bed nucleus of stria terminalis NEP
Sojourning and migrating amygdaloid neurons
Migrating and settling basal telencephalic neurons
Sojourning and migrating basal ganglionic neurons

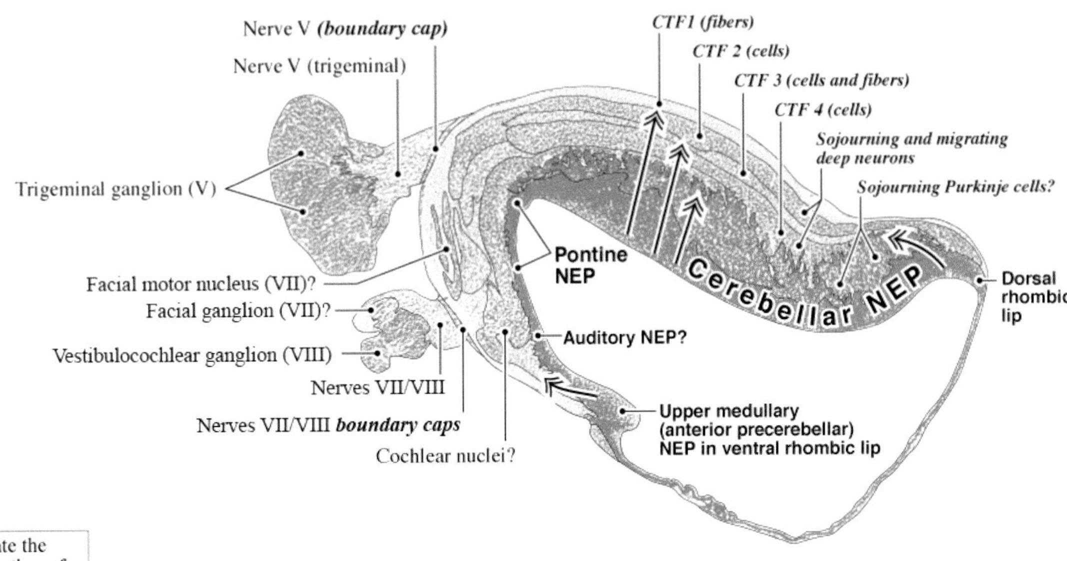

Nerve V ***(boundary cap)***
Nerve V (trigeminal)
CTF1 (fibers)
CTF 2 (cells)
CTF 3 (cells and fibers)
CTF 4 (cells)
Sojourning and migrating deep neurons
Sojourning Purkinje cells?
Trigeminal ganglion (V)
Pontine NEP
Cerebellar NEP
Dorsal rhombic lip
Facial motor nucleus (VII)?
Facial ganglion (VII)?
Auditory NEP?
Vestibulocochlear ganglion (VIII)
Nerves VII/VIII
Nerves VII/VIII ***boundary caps***
Upper medullary (anterior precerebellar) NEP in ventral rhombic lip
Cochlear nuclei?

Arrows indicate the presumed ***direction of neuron migration*** from neuroepithelial sources.

Nerve VIII (vestibulocochlear)

ABBREVIATIONS:
GEP - Glioepithelium
NEP - Neuroepithelium
CTF - Cerebellar transitional field

FONT KEY:
Germinal zone - Helvetica bold
Transient structure - Times bold italic
Permanent structure - Times Roman or **Bold**

PLATE 7A

CR 21 mm, GW 8, C6202
Sagittal, Slide 17, Section 2
Left side of brain
SKULL, MAJOR BRAIN STRUCTURES, AND VENTRICULAR DIVISIONS

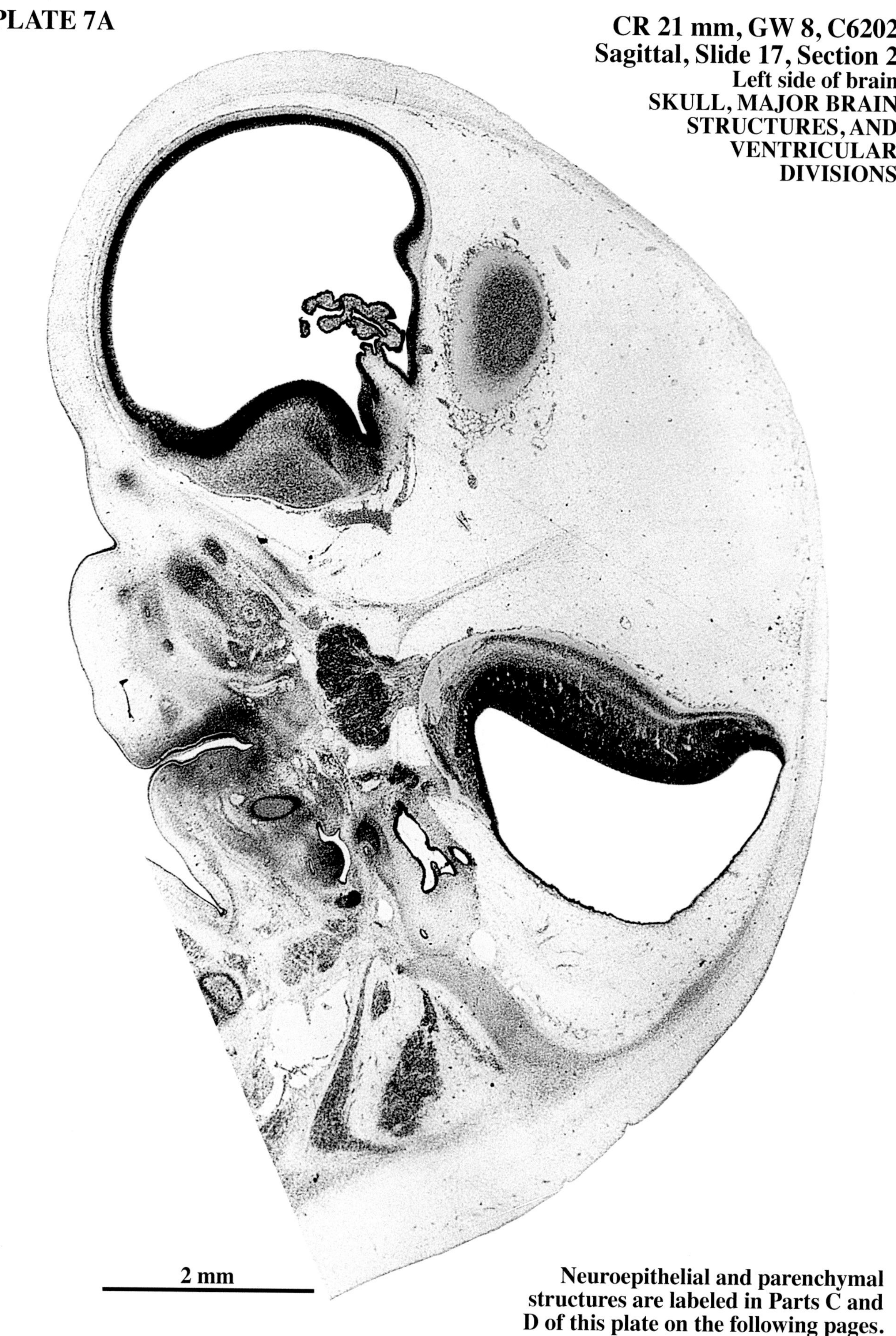
2 mm

Neuroepithelial and parenchymal structures are labeled in Parts C and D of this plate on the following pages.

PLATE 7B

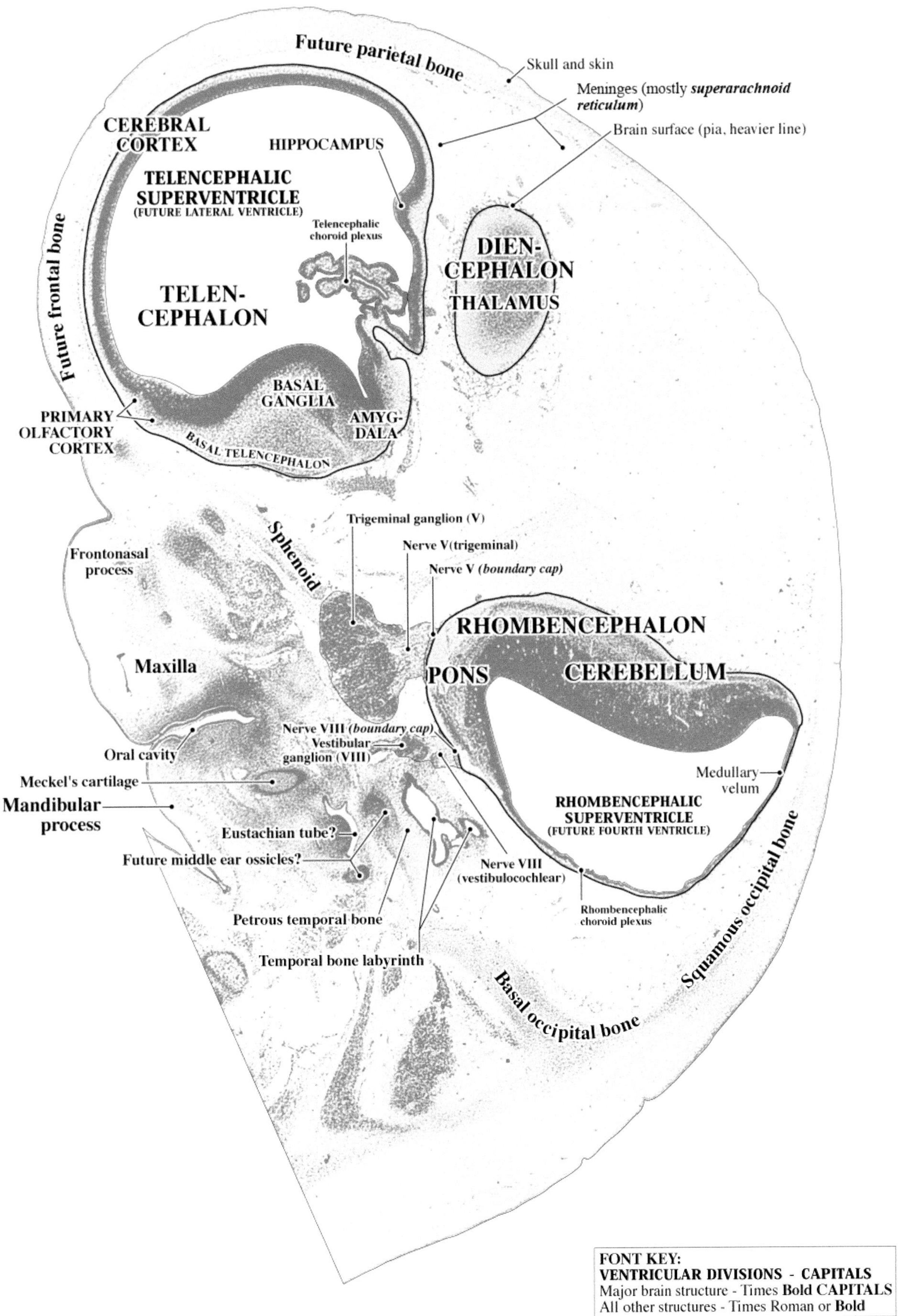
Future parietal bone
Skull and skin
Meninges (mostly ***superarachnoid reticulum***)
Brain surface (pia, heavier line)
CEREBRAL CORTEX
HIPPOCAMPUS
TELENCEPHALIC SUPERVENTRICLE
(FUTURE LATERAL VENTRICLE)
Telencephalic choroid plexus
DIEN-CEPHALON
THALAMUS
Future frontal bone
TELEN-CEPHALON
BASAL GANGLIA
AMYG-DALA
PRIMARY OLFACTORY CORTEX
BASAL TELENCEPHALON
Trigeminal ganglion (V)
Sphenoid
Nerve V (trigeminal)
Nerve V (*boundary cap*)
Frontonasal process
RHOMBENCEPHALON
Maxilla
PONS
CEREBELLUM
Nerve VIII (*boundary cap*)
Vestibular ganglion (VIII)
Oral cavity
Meckel's cartilage
Medullary velum
Mandibular process
RHOMBENCEPHALIC SUPERVENTRICLE
(FUTURE FOURTH VENTRICLE)
Eustachian tube?
Future middle ear ossicles?
Nerve VIII (vestibulocochlear)
Petrous temporal bone
Rhombencephalic choroid plexus
Temporal bone labyrinth
Squamous occipital bone
Basal occipital bone

FONT KEY:
VENTRICULAR DIVISIONS - CAPITALS
Major brain structure - Times **Bold CAPITALS**
All other structures - Times Roman or **Bold**

PLATE 7C

CR 21 mm, GW 8, C6202
Sagittal, Slide 17, Section 2
Left side of brain
NEUROEPITHELIAL AND PARENCHYMAL BRAIN STRUCTURES, PERIPHERAL GANGLIA

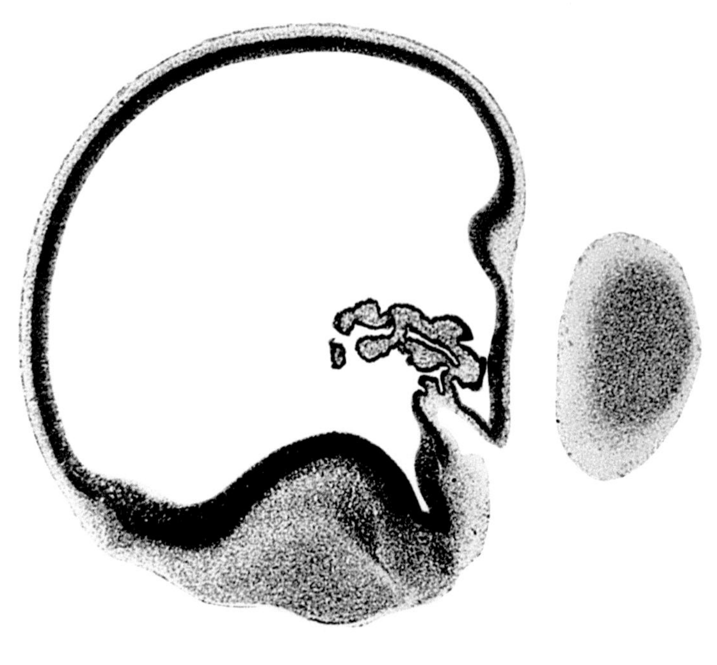

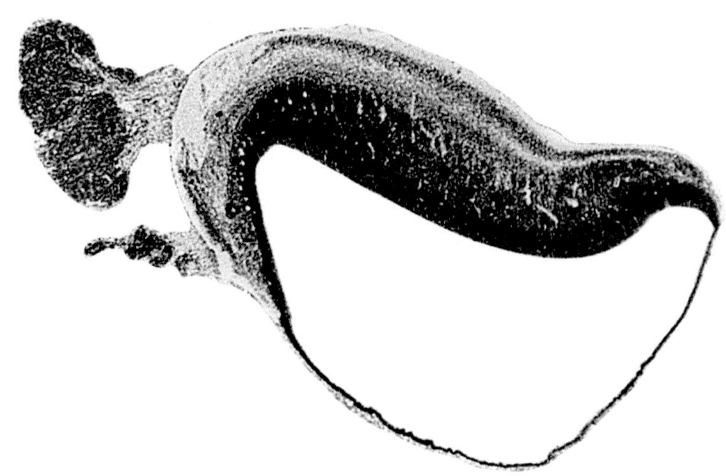

2 mm

The skull, major brain structures, and ventricular divisions are labeled in Parts A and B of this plate on the preceding pages.

PLATE 7D

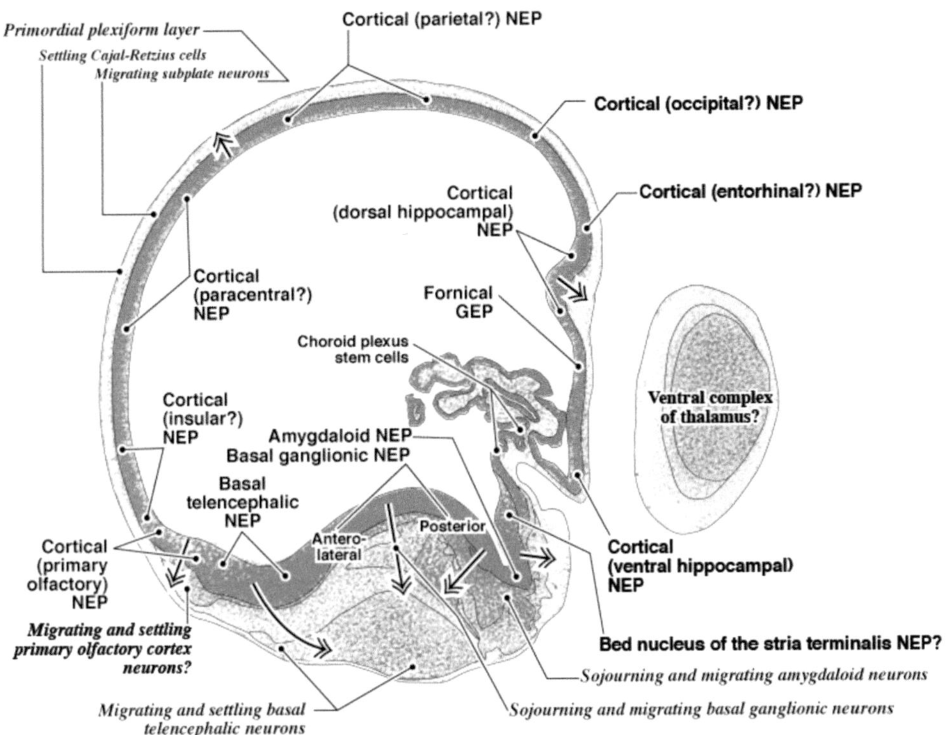

Primordial plexiform layer
Settling Cajal-Retzius cells
Migrating subplate neurons
Cortical (parietal?) NEP
Cortical (occipital?) NEP
Cortical (entorhinal?) NEP
Cortical (dorsal hippocampal) NEP
Cortical (paracentral?) NEP
Fornical GEP
Choroid plexus stem cells
Ventral complex of thalamus?
Cortical (insular?) NEP
Amygdaloid NEP
Basal ganglionic NEP
Basal telencephalic NEP
Antero-lateral
Posterior
Cortical (primary olfactory) NEP
Cortical (ventral hippocampal) NEP
Migrating and settling primary olfactory cortex neurons?
Bed nucleus of the stria terminalis NEP?
Sojourning and migrating amygdaloid neurons
Migrating and settling basal telencephalic neurons
Sojourning and migrating basal ganglionic neurons

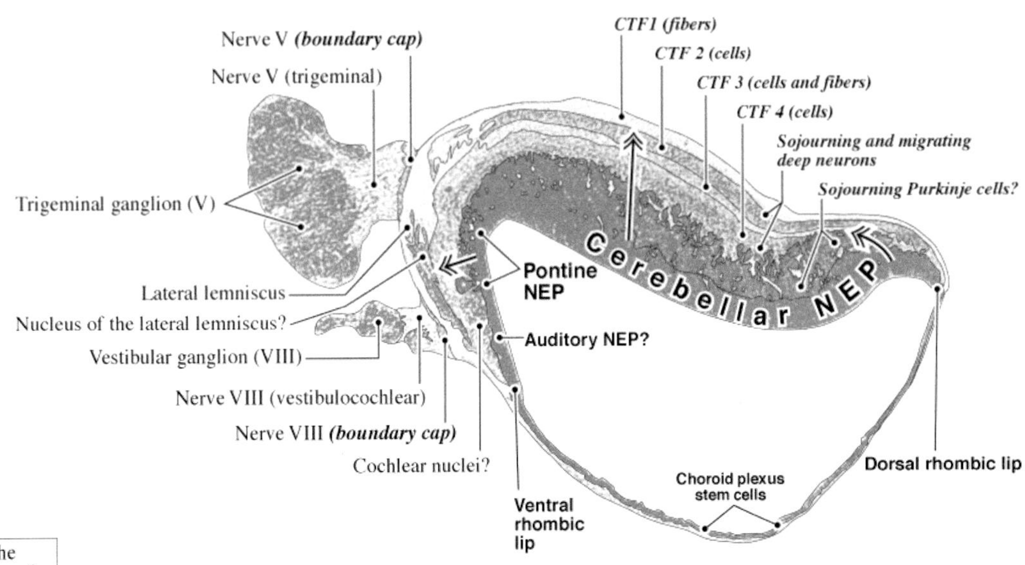

Nerve V (boundary cap)
Nerve V (trigeminal)
CTF1 (fibers)
CTF 2 (cells)
CTF 3 (cells and fibers)
CTF 4 (cells)
Sojourning and migrating deep neurons
Sojourning Purkinje cells?
Trigeminal ganglion (V)
Cerebellar NEP
Pontine NEP
Lateral lemniscus
Nucleus of the lateral lemniscus?
Auditory NEP?
Vestibular ganglion (VIII)
Nerve VIII (vestibulocochlear)
Nerve VIII (boundary cap)
Cochlear nuclei?
Dorsal rhombic lip
Choroid plexus stem cells
Ventral rhombic lip

Arrows indicate the presumed ***direction of neuron migration*** from neuroepithelial sources.

ABBREVIATIONS:
GEP - Glioepithelium
NEP - Neuroepithelium
CTF - Cerebellar transitional field

FONT KEY:
Germinal zone - Helvetica bold
Transient structure - Times bold italic
Permanent structure - Times Roman or **Bold**

PLATE 8A

CR 21 mm, GW 8, C6202, Sagittal, DORSAL CEREBRAL CORTEX

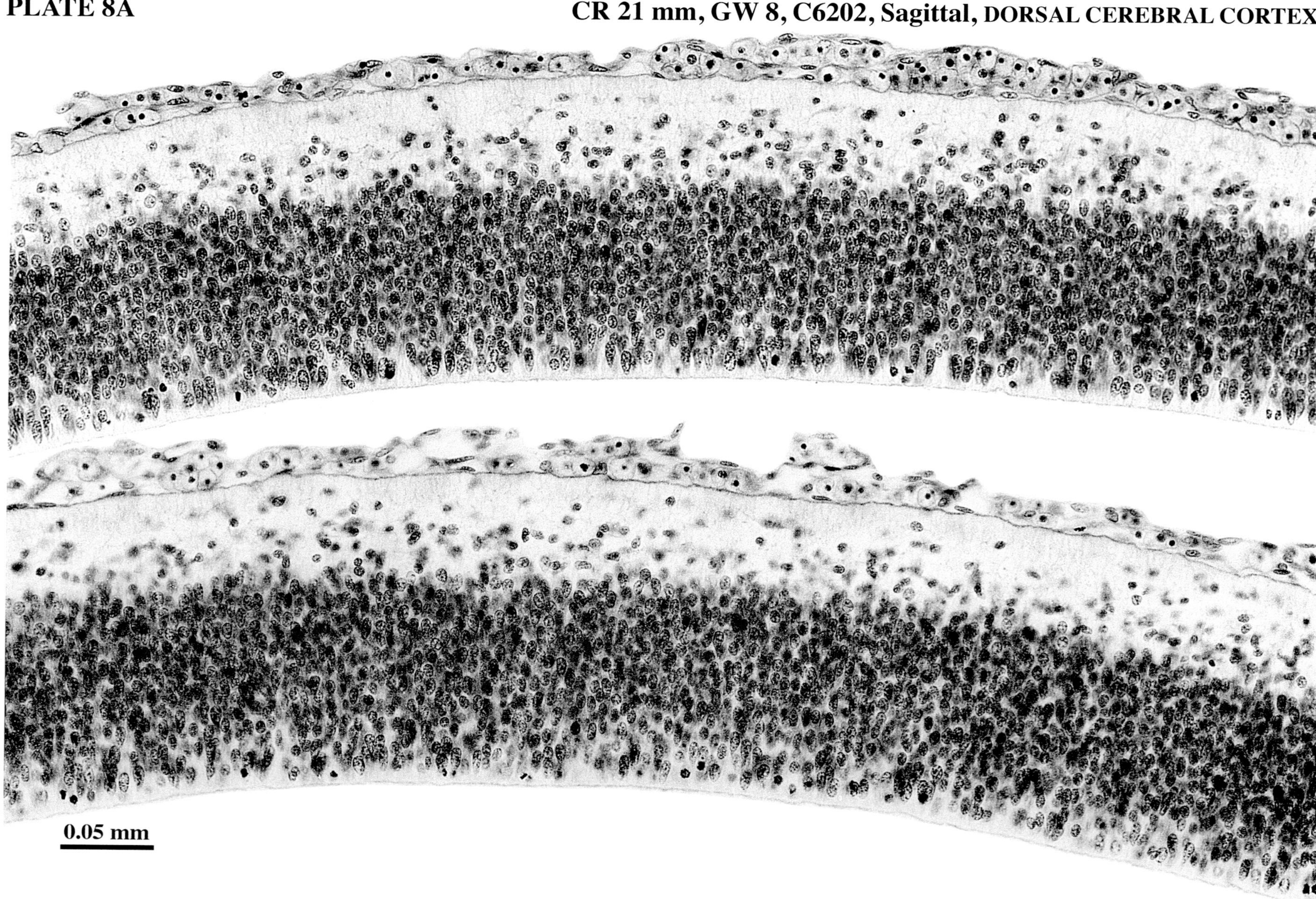
0.05 mm

PLATE 8B

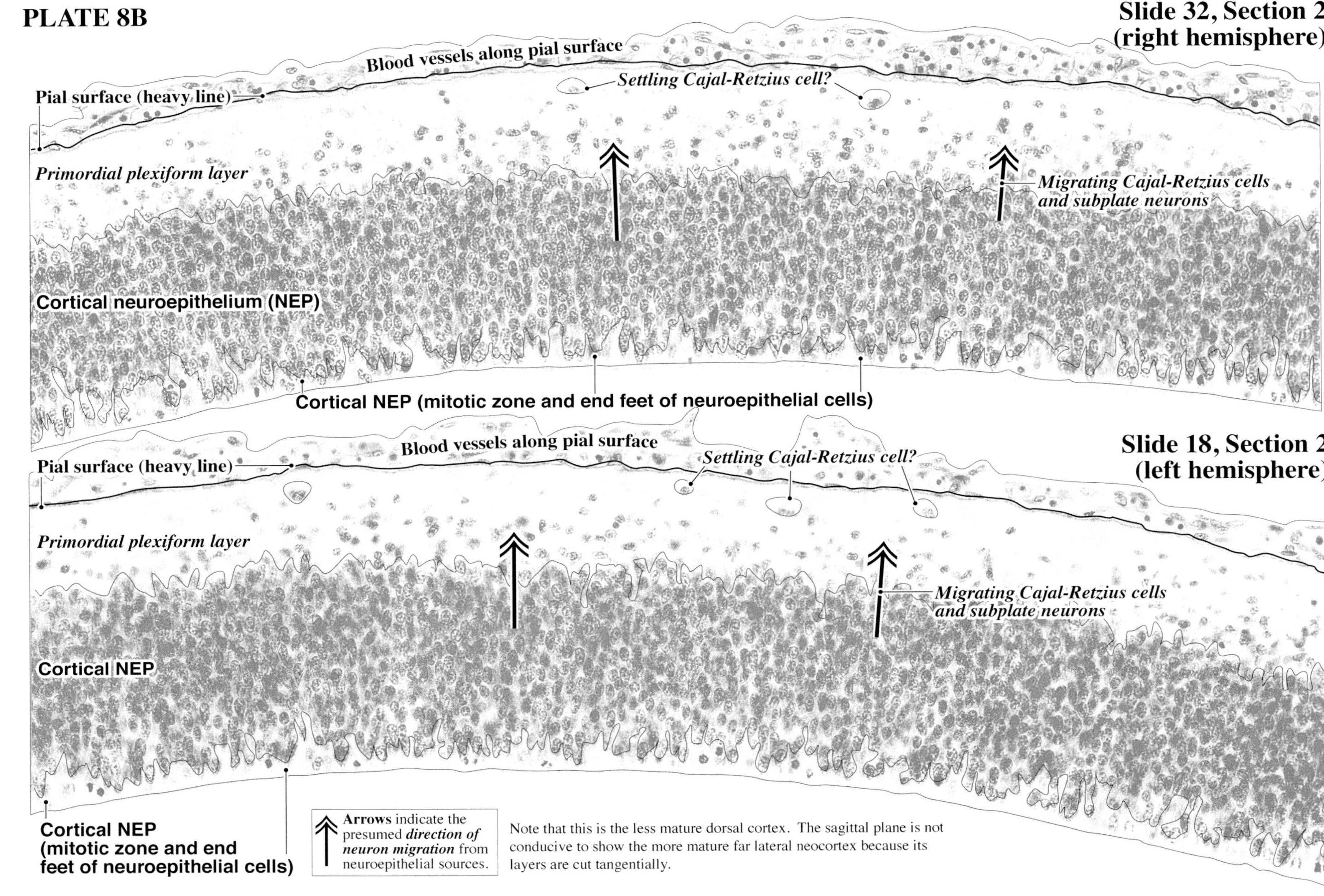
Slide 32, Section 2
(right hemisphere)
Blood vessels along pial surface
Pial surface (heavy line)
Settling Cajal-Retzius cell?
Primordial plexiform layer
Migrating Cajal-Retzius cells
and subplate neurons
Cortical neuroepithelium (NEP)
Cortical NEP (mitotic zone and end feet of neuroepithelial cells)
Slide 18, Section 2
(left hemisphere)
Blood vessels along pial surface
Pial surface (heavy line)
Settling Cajal-Retzius cell?
Primordial plexiform layer
Migrating Cajal-Retzius cells
and subplate neurons
Cortical NEP
Cortical NEP
(mitotic zone and end
feet of neuroepithelial cells)
Arrows indicate the presumed ***direction of neuron migration*** from neuroepithelial sources.

Note that this is the less mature dorsal cortex. The sagittal plane is not conducive to show the more mature far lateral neocortex because its layers are cut tangentially.

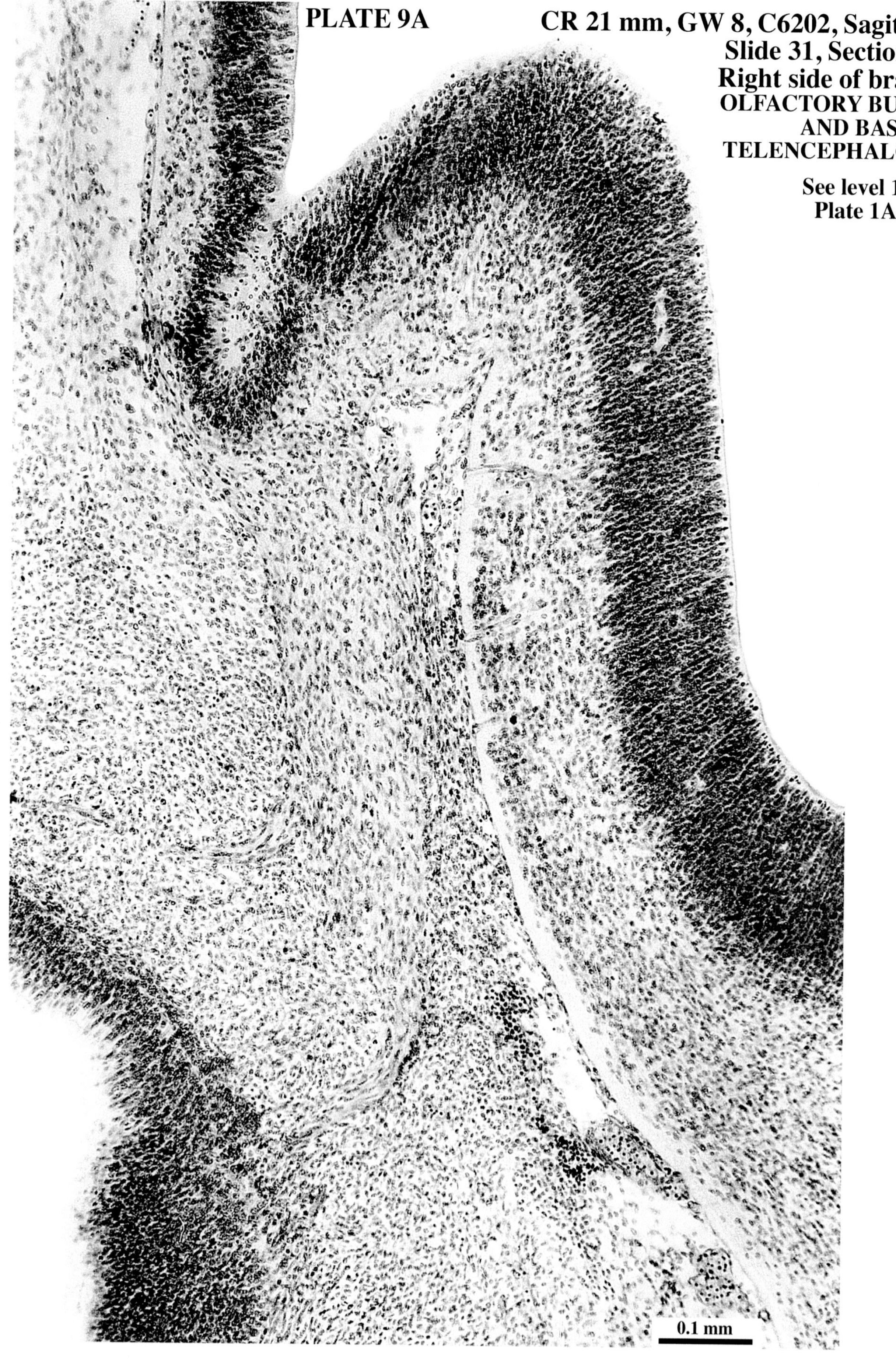
PLATE 9A
CR 21 mm, GW 8, C6202, Sagittal
Slide 31, Section 2
Right side of brain
OLFACTORY BULB
AND BASAL
TELENCEPHALON
See level 1 in
Plate 1A-D.
0.1 mm

PLATE 9B

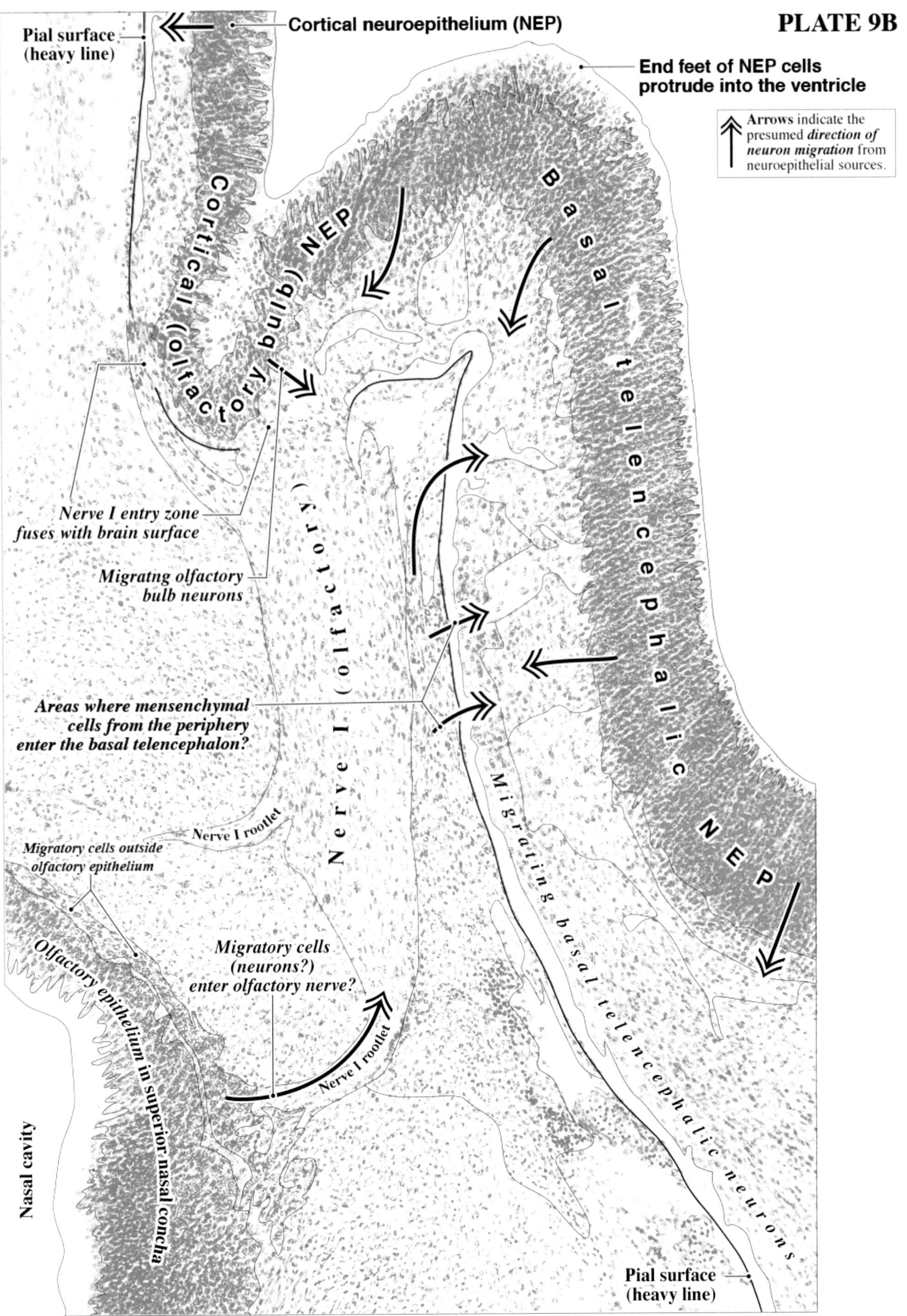
Pial surface
(heavy line)
Cortical neuroepithelium (NEP)
End feet of NEP cells
protrude into the ventricle
Arrows indicate the presumed direction of neuron migration from neuroepithelial sources.
Cortical (olfactory bulb) NEP
Basal telencephalic NEP
Nerve I entry zone
fuses with brain surface
Migratng olfactory
bulb neurons
Nerve I (olfactory)
Areas where mensenchymal
cells from the periphery
enter the basal telencephalon?
Nerve I rootlet
Migratory cells outside
olfactory epithelium
Migrating basal telencephalic neurons
Olfactory epithelium in superior nasal concha
Migratory cells
(neurons?)
enter olfactory nerve?
Nerve I rootlet
Nasal cavity
Pial surface
(heavy line)

PLATE 10A

CR 21 mm, GW 8,
C6202, Sagittal
Slide 18, Section 2
Left side of brain
TELENCEPHALON
AND DIENCEPHALON

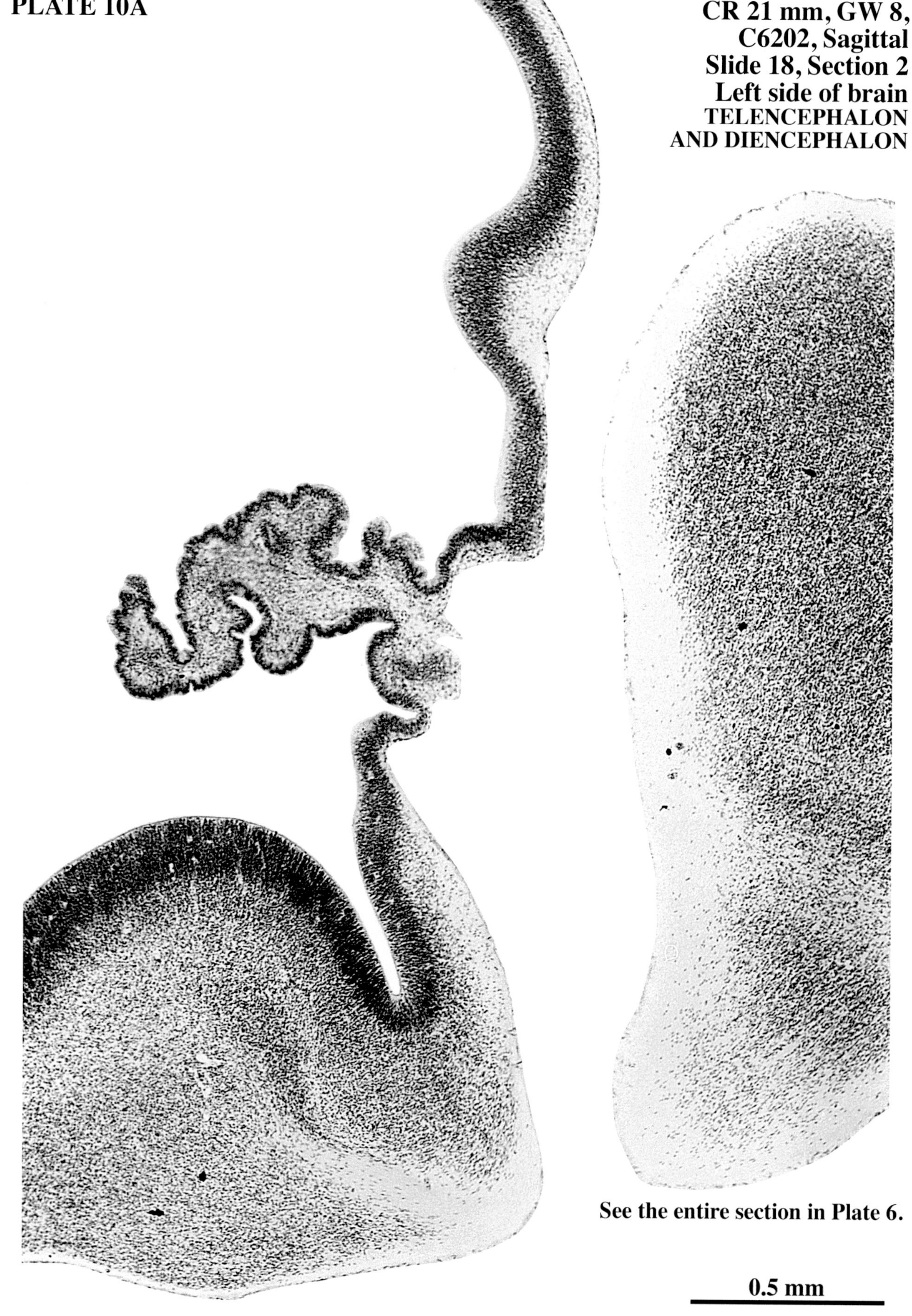

See the entire section in Plate 6.

0.5 mm

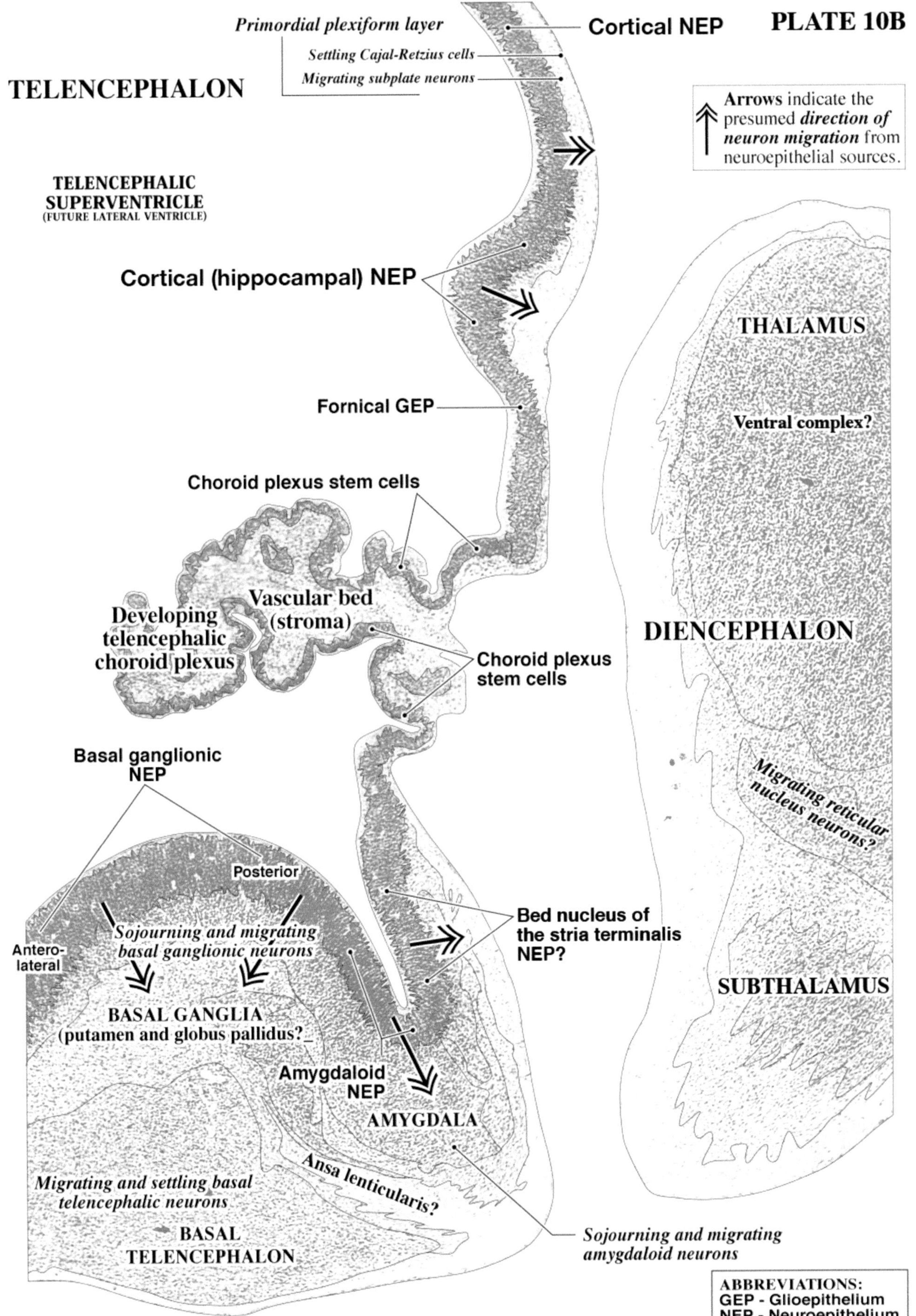
PLATE 10B
TELENCEPHALON
Primordial plexiform layer
Settling Cajal-Retzius cells
Migrating subplate neurons
Cortical NEP
Arrows indicate the presumed direction of neuron migration from neuroepithelial sources.
TELENCEPHALIC SUPERVENTRICLE
(FUTURE LATERAL VENTRICLE)
Cortical (hippocampal) NEP
THALAMUS
Ventral complex?
Fornical GEP
Choroid plexus stem cells
Vascular bed (stroma)
Developing telencephalic choroid plexus
DIENCEPHALON
Choroid plexus stem cells
Basal ganglionic NEP
Migrating reticular nucleus neurons?
Posterior
Antero-lateral
Sojourning and migrating basal ganglionic neurons
Bed nucleus of the stria terminalis NEP?
SUBTHALAMUS
BASAL GANGLIA
(putamen and globus pallidus?_
Amygdaloid NEP
AMYGDALA
Ansa lenticularis?
Migrating and settling basal telencephalic neurons
BASAL TELENCEPHALON
Sojourning and migrating amygdaloid neurons
ABBREVIATIONS:
GEP - Glioepithelium
NEP - Neuroepithelium

PLATE 11A

**CR 21 mm, GW 8, C6202, Sagittal,
Slide 28, Section 3
HYPOTHALAMUS**

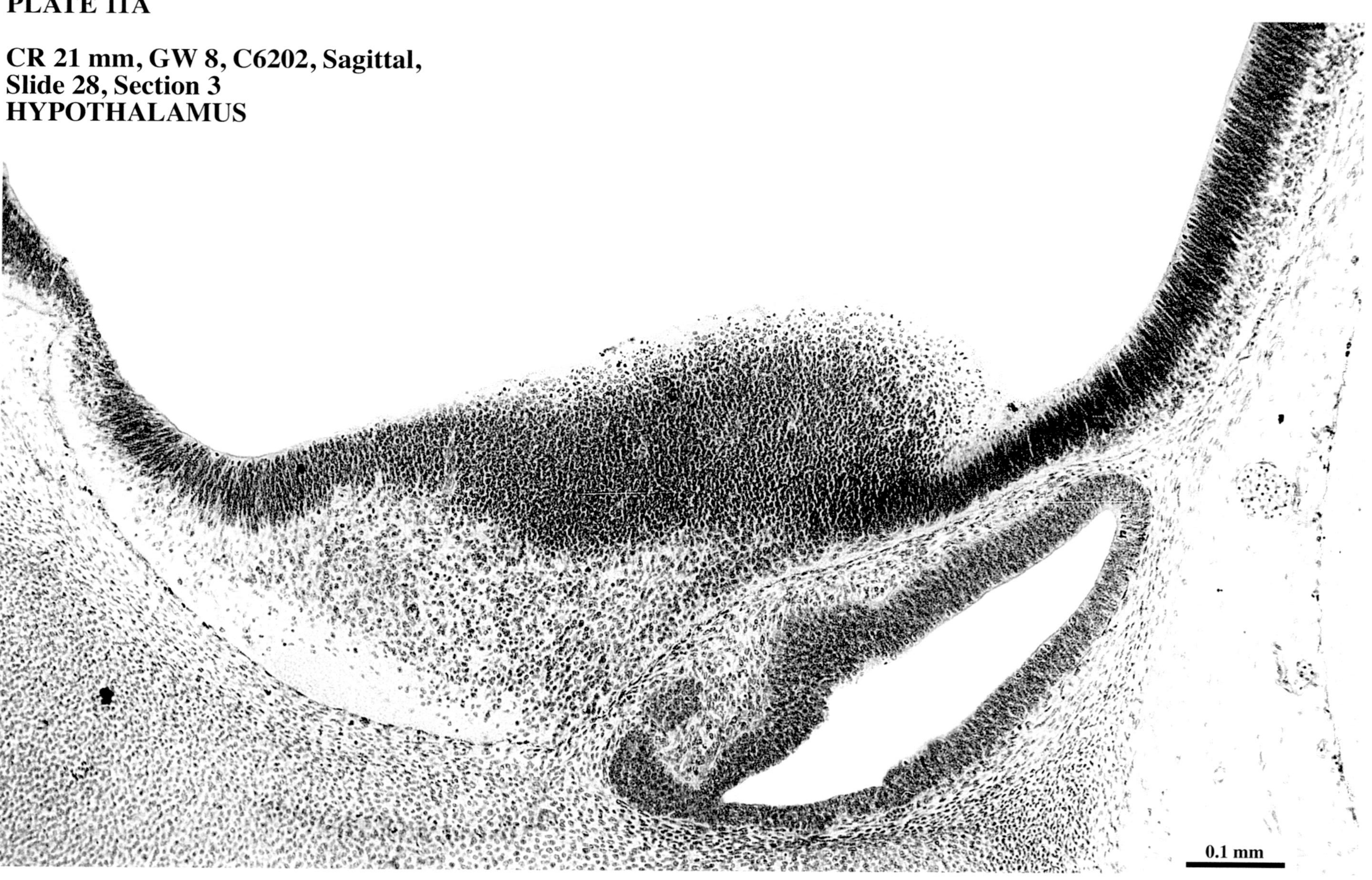
0.1 mm

See nearby complete sections in Plates 1 and 2.

PLATE 11B

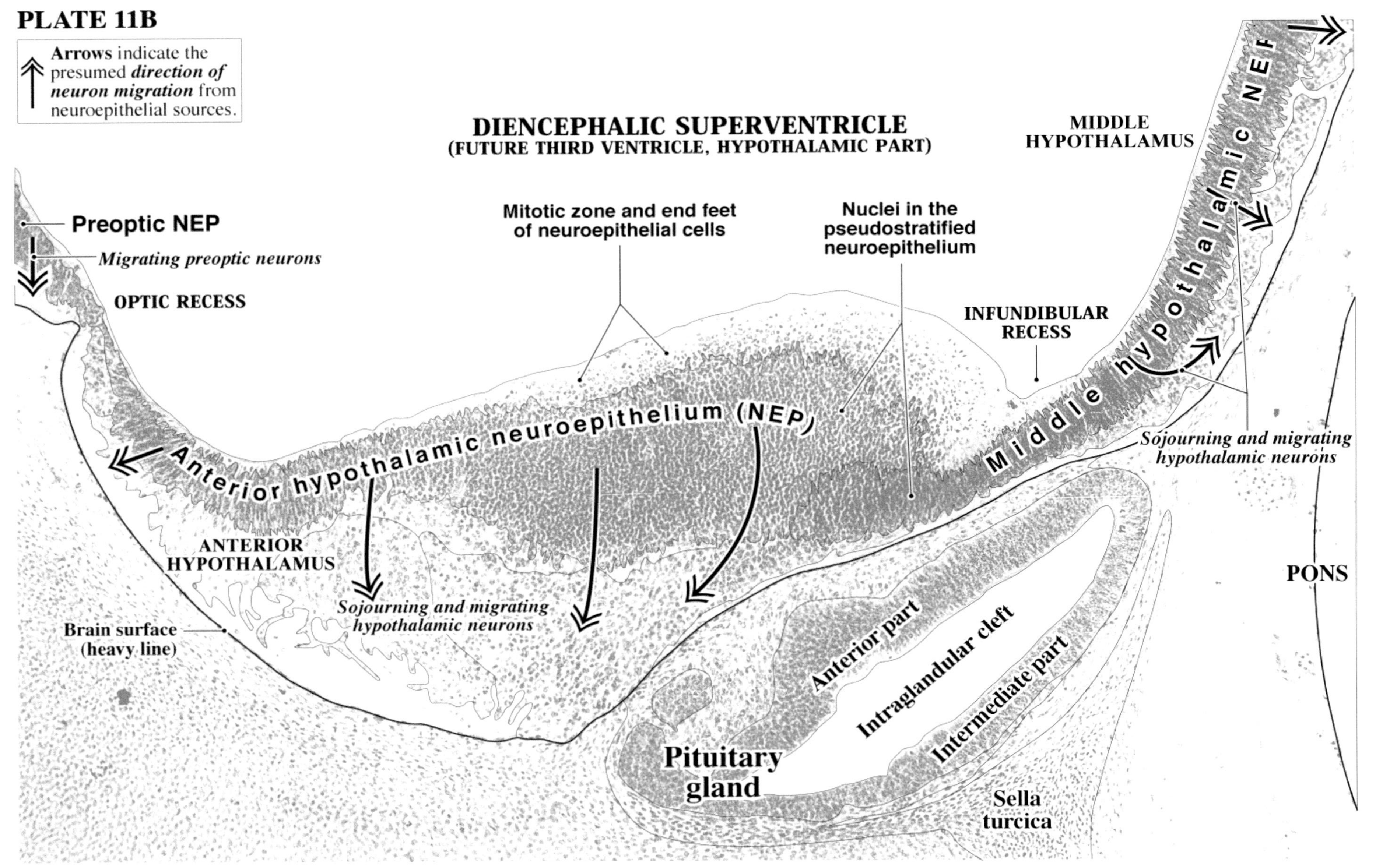
Arrows indicate the presumed direction of neuron migration from neuroepithelial sources.
DIENCEPHALIC SUPERVENTRICLE
(FUTURE THIRD VENTRICLE, HYPOTHALAMIC PART)
MIDDLE HYPOTHALAMUS
Preoptic NEP
Migrating preoptic neurons
OPTIC RECESS
Mitotic zone and end feet of neuroepithelial cells
Nuclei in the pseudostratified neuroepithelium
INFUNDIBULAR RECESS
Middle hypothalamic NEP
Anterior hypothalamic neuroepithelium (NEP)
Sojourning and migrating hypothalamic neurons
ANTERIOR HYPOTHALAMUS
Sojourning and migrating hypothalamic neurons
Brain surface (heavy line)
PONS
Anterior part
Intraglandular cleft
Intermediate part
Pituitary gland
Sella turcica

PLATE 12A

CR 21 mm, GW 8, C6202, Sagittal,
Slide 28, Section 3
MIDBRAIN
TEGMENTUM

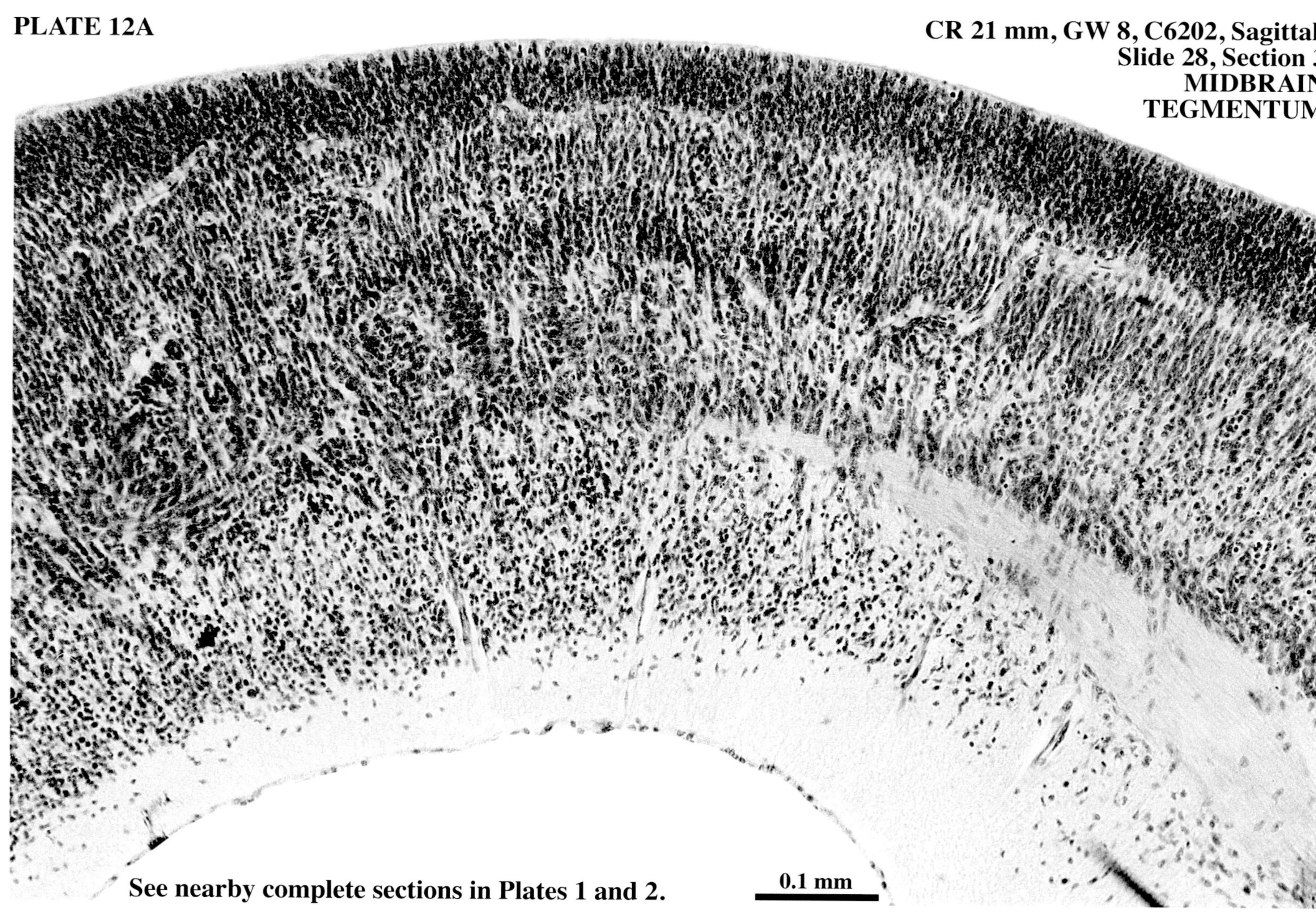
0.1 mm

See nearby complete sections in Plates 1 and 2.

PLATE 12B

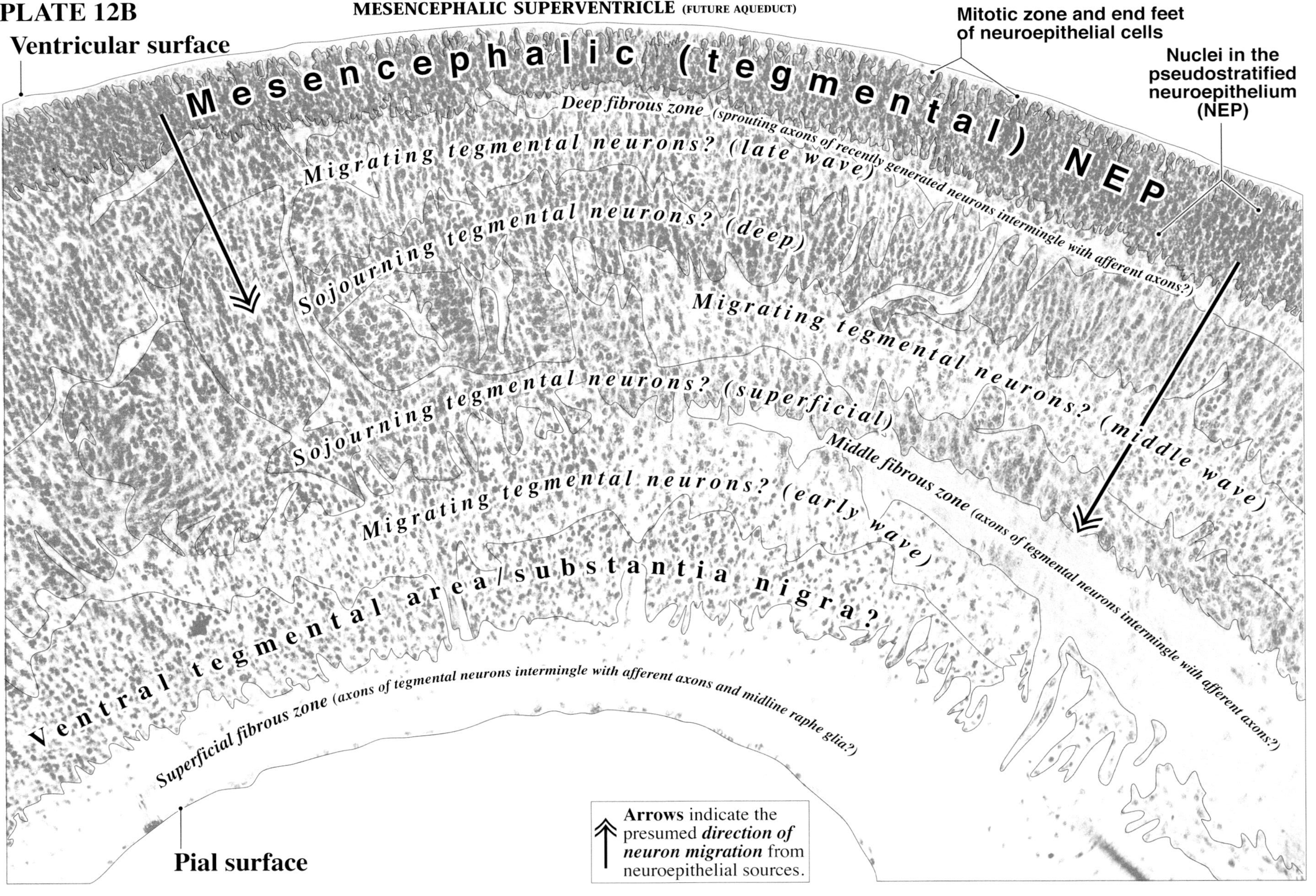
MESENCEPHALIC SUPERVENTRICLE (FUTURE AQUEDUCT)
Ventricular surface
Mitotic zone and end feet of neuroepithelial cells
Nuclei in the pseudostratified neuroepithelium (NEP)
Mesencephalic (tegmental) NEP
Deep fibrous zone (sprouting axons of recently generated neurons intermingle with afferent axons?)
Migrating tegmental neurons? (late wave)
Sojourning tegmental neurons? (deep)
Migrating tegmental neurons? (middle wave)
Sojourning tegmental neurons? (superficial)
Middle fibrous zone (axons of tegmental neurons intermingle with afferent axons?)
Migrating tegmental neurons? (early wave)
Ventral tegmental area/substantia nigra?
Superficial fibrous zone (axons of tegmental neurons intermingle with afferent axons and midline raphe glia?)
Pial surface
Arrows indicate the presumed direction of neuron migration from neuroepithelial sources.

PLATE 13A

CR 21 mm, GW 8, C6202, Sagittal,
CEREBELLUM

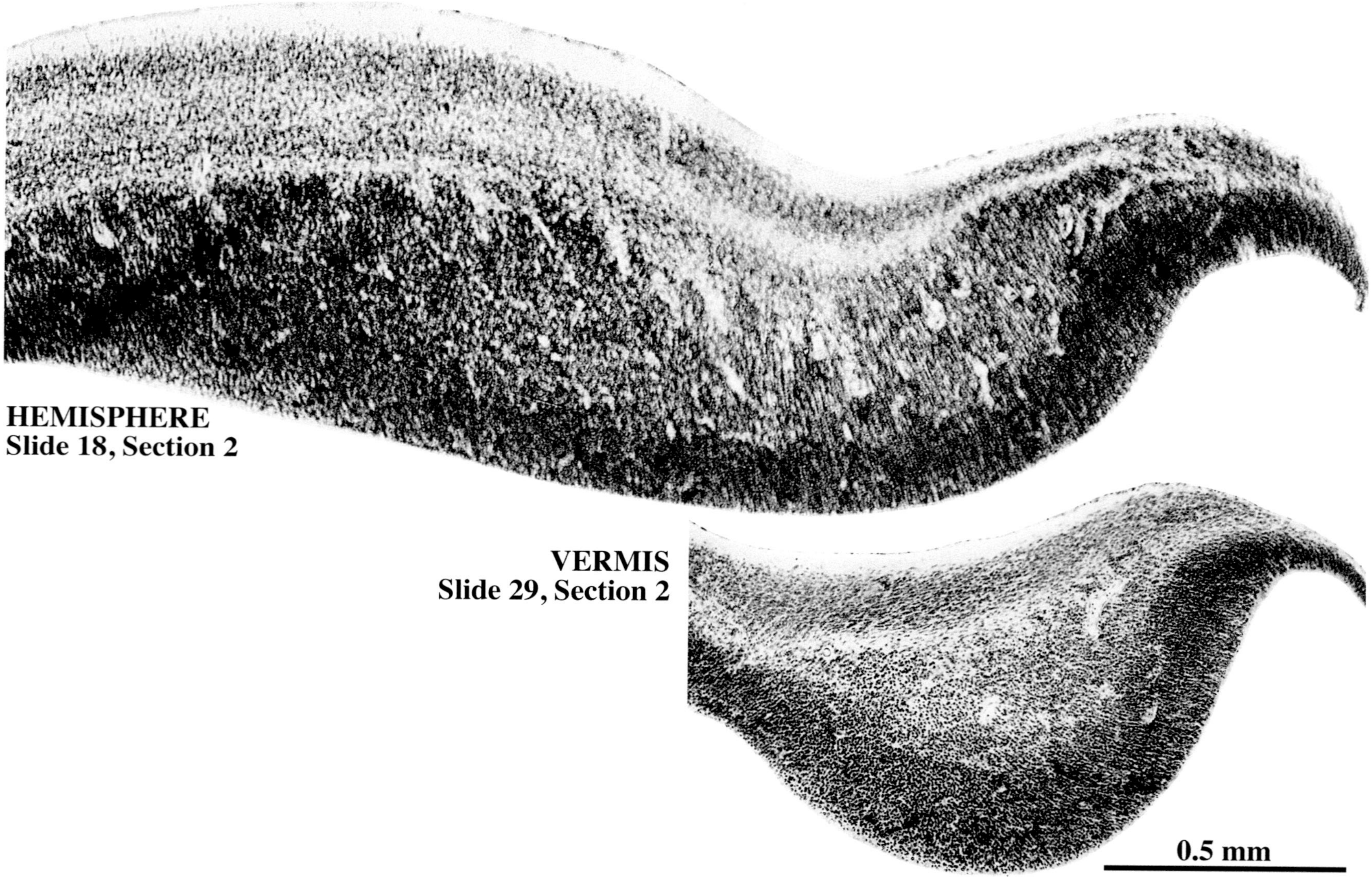

HEMISPHERE
Slide 18, Section 2

VERMIS
Slide 29, Section 2

0.5 mm

See nearby complete sections in Plates 1 and 6.

PLATE 13B

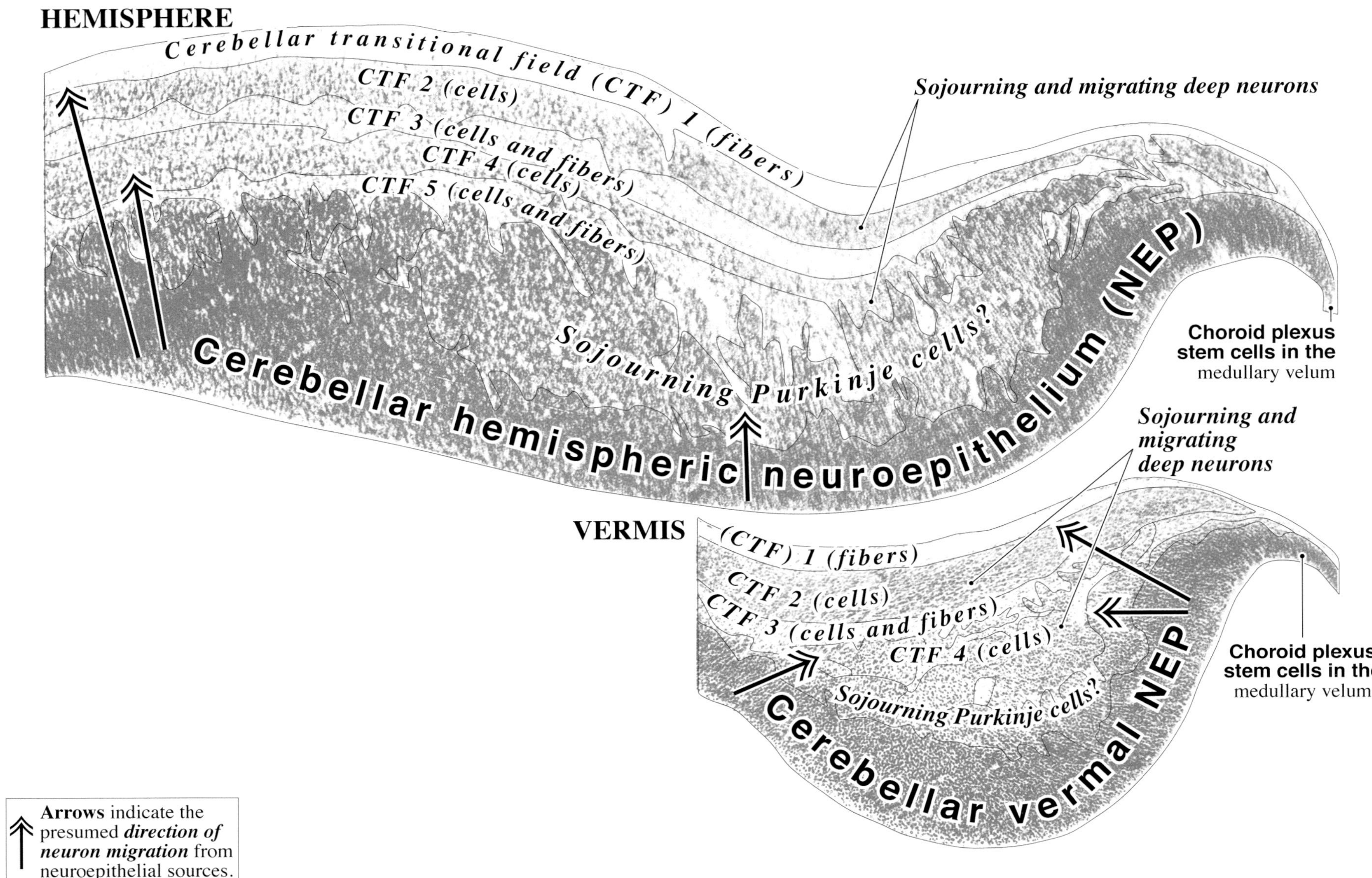
HEMISPHERE
Cerebellar transitional field (CTF) 1 (fibers)
CTF 2 (cells)
CTF 3 (cells and fibers)
CTF 4 (cells)
CTF 5 (cells and fibers)
Sojourning and migrating deep neurons
Sojourning Purkinje cells?
Cerebellar hemispheric neuroepithelium (NEP)
Choroid plexus stem cells in the medullary velum
VERMIS
(CTF) 1 (fibers)
CTF 2 (cells)
CTF 3 (cells and fibers)
CTF 4 (cells)
Sojourning and migrating deep neurons
Sojourning Purkinje cells?
Cerebellar vermal NEP
Choroid plexus stem cells in the medullary velum
Arrows indicate the presumed ***direction of neuron migration*** from neuroepithelial sources.

PLATE 14A

CR 21 mm, GW 8, C6202, Sagittal,
Slide 18, Section 2
PONS AND PERIPHERAL GANGLIA

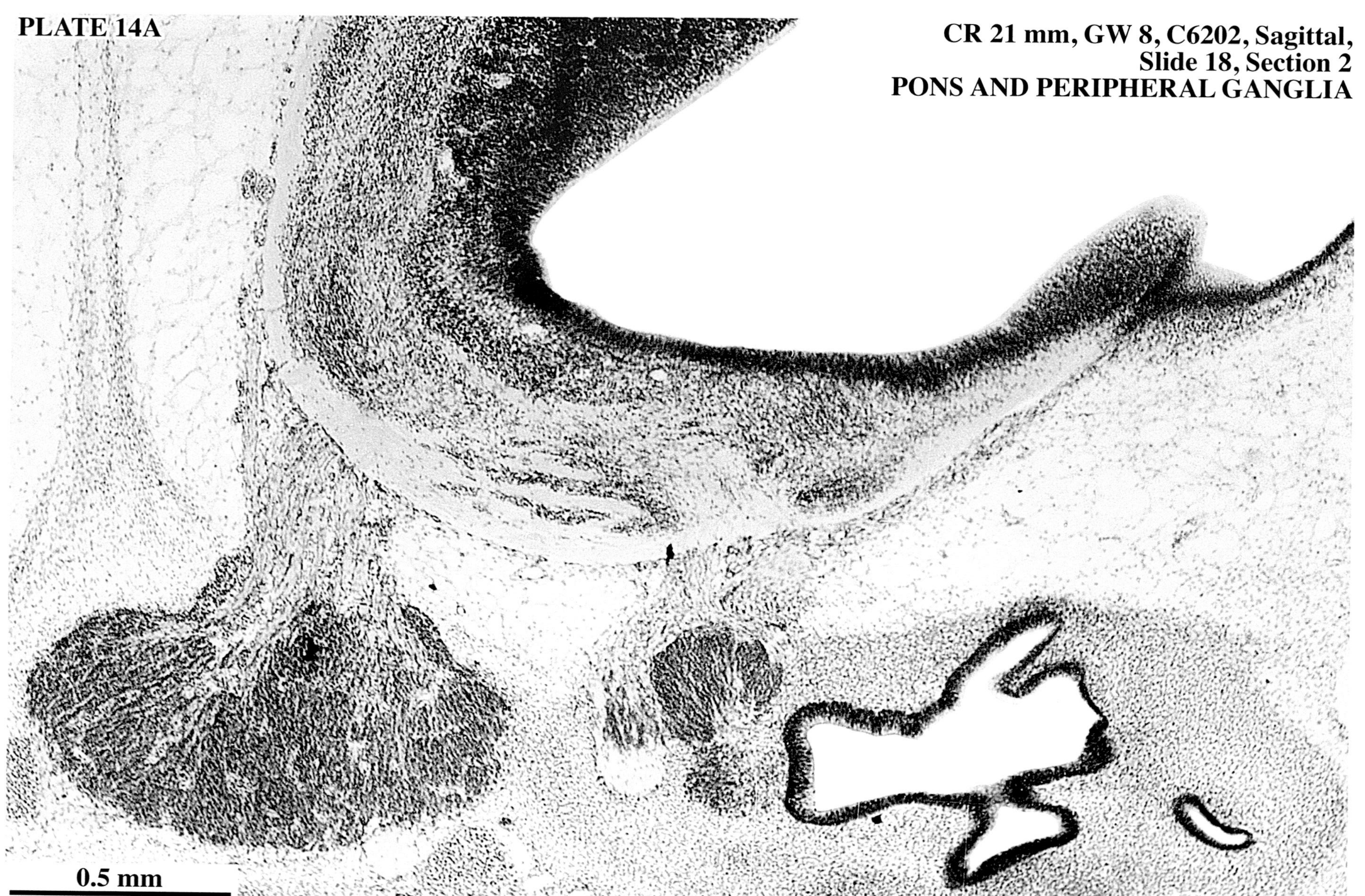
0.5 mm

See the entire section in Plate 6.

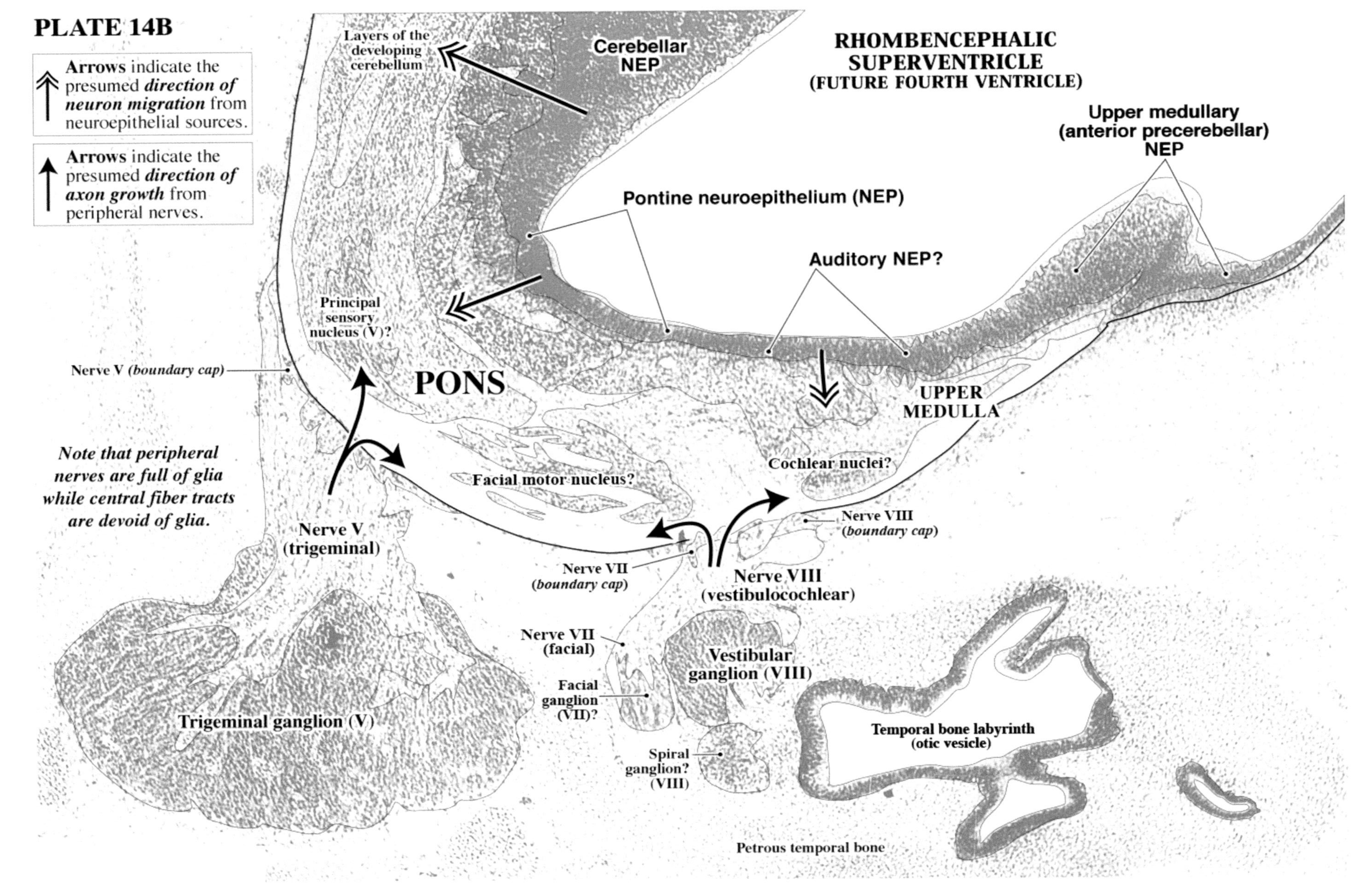

PLATE 14B
Arrows indicate the presumed *direction of neuron migration* from neuroepithelial sources.
Arrows indicate the presumed *direction of axon growth* from peripheral nerves.
Layers of the developing cerebellum
Cerebellar NEP
RHOMBENCEPHALIC SUPERVENTRICLE (FUTURE FOURTH VENTRICLE)
Upper medullary (anterior precerebellar) NEP
Pontine neuroepithelium (NEP)
Auditory NEP?
Principal sensory nucleus (V)?
Nerve V (*boundary cap*)
PONS
UPPER MEDULLA
Note that peripheral nerves are full of glia while central fiber tracts are devoid of glia.
Facial motor nucleus?
Cochlear nuclei?
Nerve VIII (*boundary cap*)
Nerve V (trigeminal)
Nerve VII (*boundary cap*)
Nerve VIII (vestibulocochlear)
Nerve VII (facial)
Vestibular ganglion (VIII)
Facial ganglion (VII)?
Trigeminal ganglion (V)
Temporal bone labyrinth (otic vesicle)
Spiral ganglion? (VIII)
Petrous temporal bone

PLATE 15A

CR 21 mm, GW 8, C6202, Sagittal, MIDLINE RAPHE GLIAL STRUCTURE

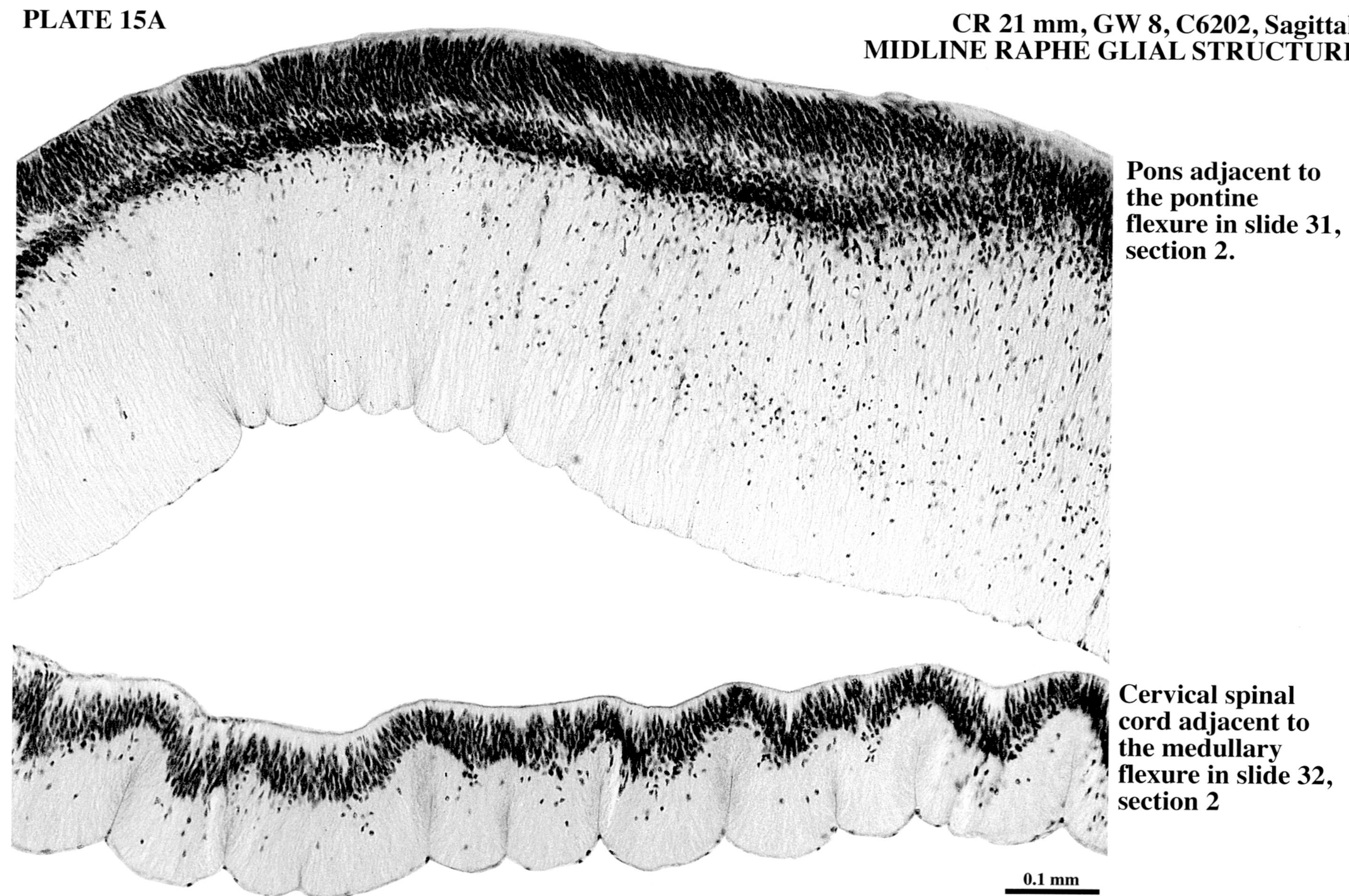
Pons adjacent to the pontine flexure in slide 31, section 2.
Cervical spinal cord adjacent to the medullary flexure in slide 32, section 2
0.1 mm

Note that both of these sections are medial to Plate 1.

PLATE 15B

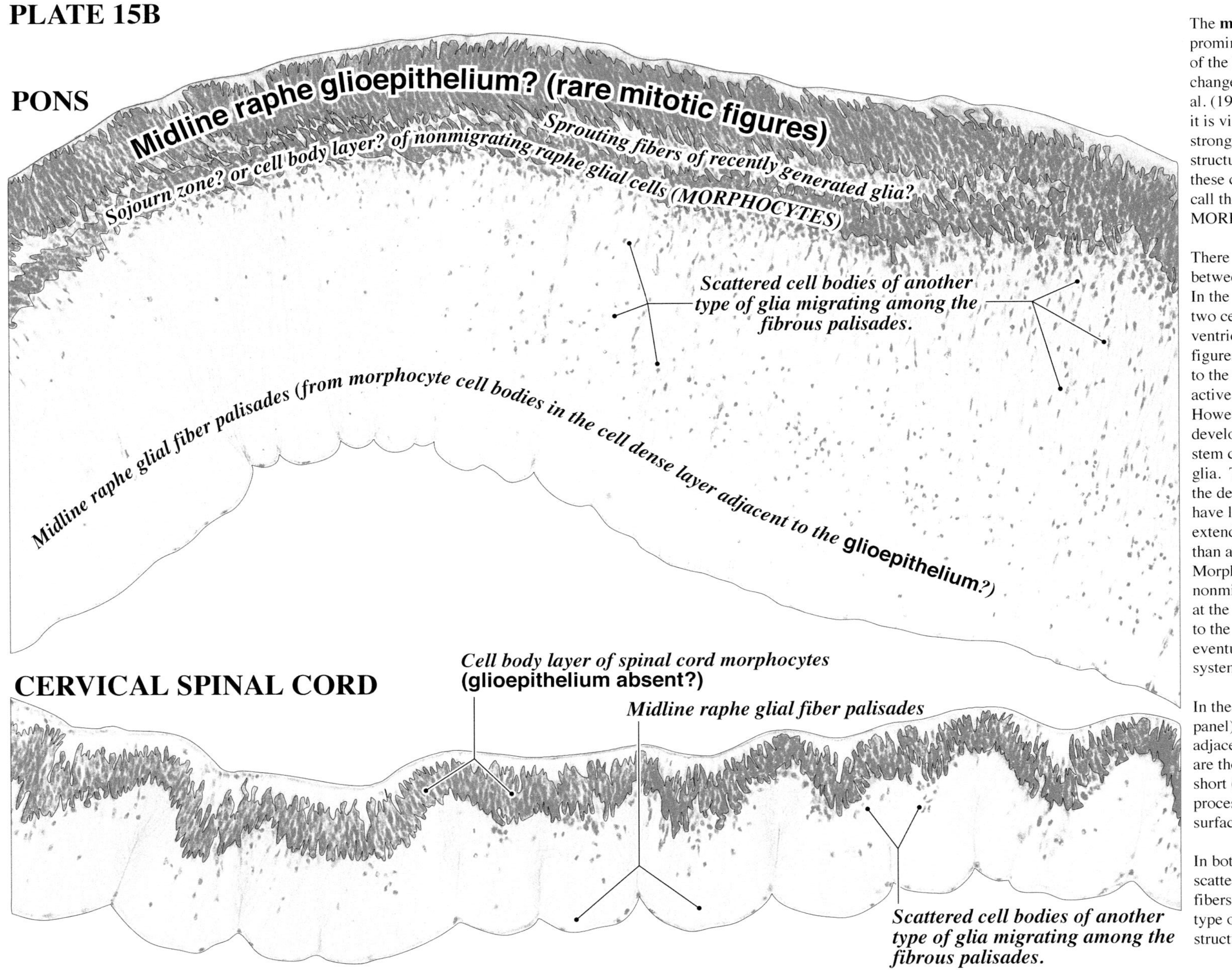
PONS
Midline raphe glioepithelium? (rare mitotic figures)
Sprouting fibers of recently generated glia?
Sojourn zone? or cell body layer? of nonmigrating raphe glial cells (MORPHOCYTES)
Scattered cell bodies of another type of glia migrating among the fibrous palisades.
Midline raphe glial fiber palisades (from morphocyte cell bodies in the cell dense layer adjacent to the glioepithelium?)
CERVICAL SPINAL CORD
Cell body layer of spinal cord morphocytes (glioepithelium absent?)
Midline raphe glial fiber palisades
Scattered cell bodies of another type of glia migrating among the fibrous palisades.

The **midline raphe glial system** is prominent in regions where the shape of the brain and spinal cord sharply change curvature. Van Hrtesveldt et al. (1986) described this in rats, and it is virtually identical in man. The strong fibrous palisades may provide structural stability in the region of these curvatures. Consequently, we call the specialized glia MORPHOCYTES.

There is structural variability between the pons and spinal cord. In the PONS (top panel) there are two cell-dense layers near the ventricular surface. Since mitotic figures are rare in the layer adjacent to the ventricle, it may not be an active germinal zone generating glia. However, at earlier stages of development, this layer is full of stem cells generating midline raphe glia. The second layer is most likely the densely packed cell bodies that have long fibrous processes extending to the pial surface rather than a premigratory sojourn zone. Morphocytes may be predominantly nonmigratory cells that differentiate at the site of their generation (similar to the ependymal cells that eventually line the ventricular system).

In the SPINAL CORD (bottom panel) there is one cell-dense layer adjacent to the central canal. These are the cell bodies with relatively short (compared to the pons) fibrous processes that extend to the pial surface.

In both regions, there are widely scattered cells dispersed between the fibers. These are most likely another type of glial cell that do not play a structural role in the raphe system.

PART III: C966 CR 23 mm (GW 8.4) Frontal/Horizontal

This is specimen number 966 in the Carnegie collection, designated here as C966. A normal female fetus with a crown rump length (CR) of 23 mm was collected in 1914. The fetus is estimated to be at gestational week (GW) 8.4. The entire fetus was fixed in bichloric acetic acid, embedded in celloidin, cut transversely in 40-μm sections, and was stained with aluminum cochineal. The histological preservation of this specimen is excellent, and the sections are cut nearly perfectly bilaterally. Several years ago, an excellent 3-D reconstruction of the brain and upper cervical spinal cord was done by piecing together cardboard cutouts of the brain outlines in each section and then gluing them together; the rhombencephalic superventricle was not included in that reconstruction. A photograph of that model (which is still part of the Carnegie Collection today) shows us the exact location and cutting plane of C966's sections (**Fig. 21**). Like most of the specimens in this volume, the sections are not cut exactly in one plane, but C966's sections are much closer to the frontal than the horizontal plane. Photographs of 21 sections of the brain in the head are shown in **Plates 16-36**. Our computer-aided 3D reconstructions of the brain, the ventricles (including the rhombencephalic superventricle), and selected neuroepithelial components are shown in **Figures 22-31** following the methods explained in **Part I**.

The superventricles are large in the centers of all brain structures, especially in the telencephalon and rhombencephalon. The parenchyma is thick and bordered by a thin neuroepithelium (NEP) that is transitioning to an ependymal layer in the medulla, pons, and midbrain tegmentum, where most neurons have already been generated. There are layers of dense cells adjacent to the lateral pontine NEP, where vestibular nuclear neurons and trigeminal nuclear neurons may be sojourning prior to migration and settling. There is a large accumulation of presumptive facial motor neurons in the pons that have migrated toward the incoming facial nerve.

The precerebellar NEP in the medulla is thick and generating precerebellar neurons. Many inferior olive neurons have already settled in the medulla, confirming the neurogenetic data in rats that the inferior olive contains the oldest precerebellar neurons, but many are still migrating in the parenchyma (posterior intramural migratory stream). Neurons are now migrating in a posterior extramural migratory stream that will cross the midline and settle in the contralateral external cuneate and lateral reticular nuclei. The cerebellar NEP itself is thick and difficult to distinguish from an adjacent dense sojourn zone in the cerebellar parenchyma, called the cerebellar transitional field (CTF) 6. The remaining CTF has alternating layers of cells and fibers (CTF1-5). The external germinal layer is completely absent. If one can extrapolate from data on cerebellar neurogenesis in rats, the human cerebellar NEP by GW8.4 has generated all of the deep neurons and most of the Purkinje cells are sojourning in CTF6. However, the enormous size of the human cerebellum may have extended neurogenetic periods compared to rats.

Both the mesencephalic tegmental NEP and the isthmal NEP are nearly the same thickness as the pontine and medullary NEPs, but dense sojourn zones of young neurons are more obvious in the adjacent parenchyma. The superficial border of a thick mesencephalic tectal NEP is difficult to distinguish from dense accumulations of young neurons sequestered in the basal tectal NEPs extending into a thin parenchyma. Many neurons in both the superior and inferior colliculi are currently being generated.

The diencephalic NEP is very thick as most NEPs are in their neurogenetic phases. The subthalamic NEPs are thinnest, and most neurons there have already been generated. The thin parenchyma in the thalamus is filled with dense zones of sojourning and migrating neurons. More neurons are in the parenchyma of the hypothalamus. An open channel marks the area where fibers will invade the internal capsule, and some pioneer axons are accumulating in inner and outer fibrous layers in the thalamic parenchyma.

The telencephalic NEP is thick in all cortical areas; this is the beginning of a prolonged neurogenetic phase in all cortical areas. There is little to no parenchyma outside the cortical NEP, rather, a thin primordial plexiform layer that contains the oldest cortical neurons (Cajal-Retzius cells) and subplate neurons; there is a primordial cortical plate in a central ventrolateral area that probably marks the future insular area. The oldest basal telencephalic and basal ganglionic neurons are settling in a thick parenchyma. Some neurons in the septum are still to be generated, and neurogenesis in the striatum has barely begun because the striatal subventricular zone is just being established.

CR 23 mm, C966, GW 8.4 FRONTAL/HORIZONTAL SECTION PLANES

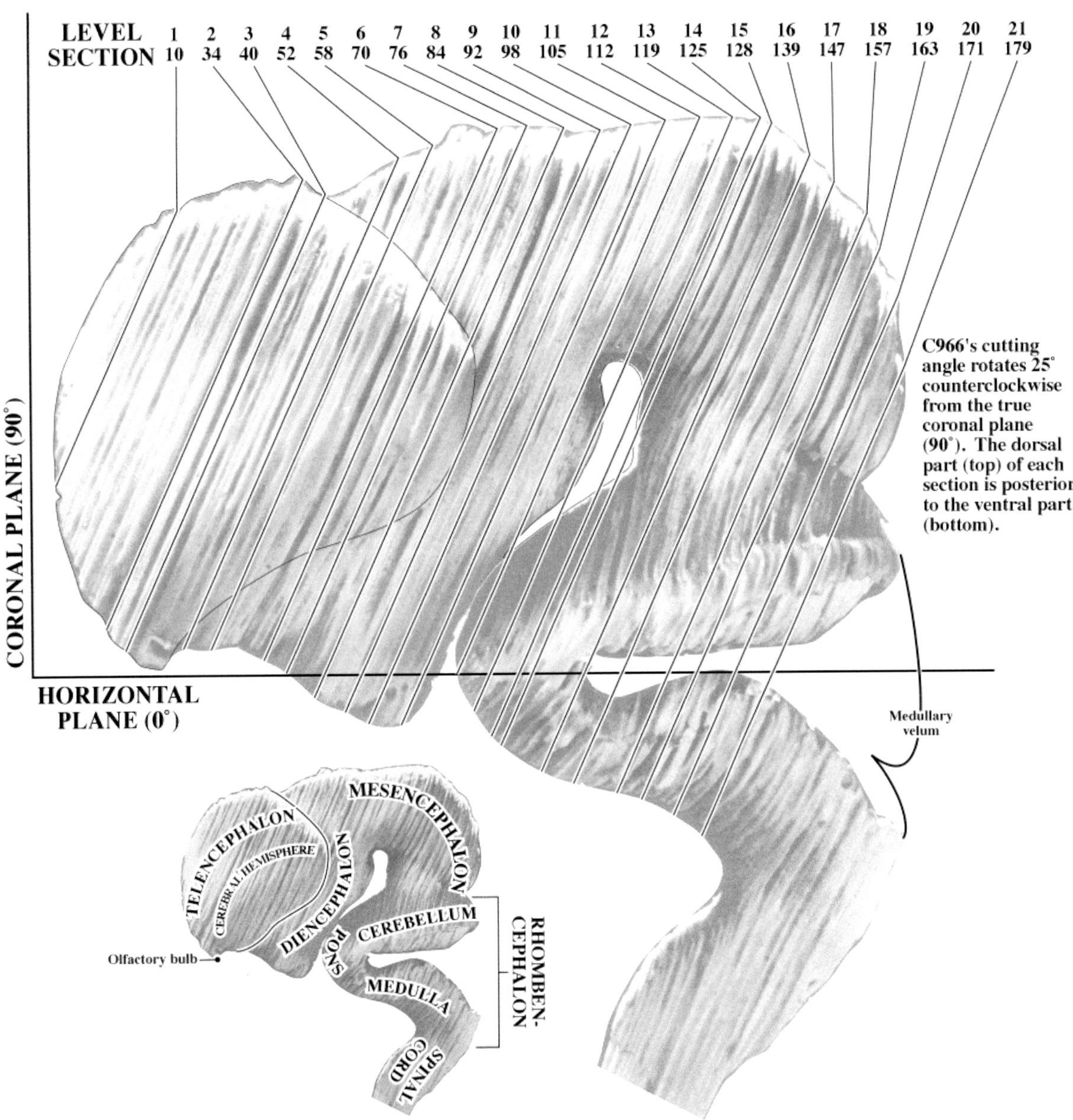
LEVEL 1 2 3 4 5 6 7 8 9 10 11 12 13 14 15 16 17 18 19 20 21
SECTION 10 34 40 52 58 70 76 84 92 98 105 112 119 125 128 139 147 157 163 171 179
C966's cutting angle rotates 25° counterclockwise from the true coronal plane (90°). The dorsal part (top) of each section is posterior to the ventral part (bottom).
CORONAL PLANE (90°)
HORIZONTAL PLANE (0°)
Medullary velum
TELENCEPHALON
CEREBRAL HEMISPHERE
MESENCEPHALON
DIENCEPHALON
CEREBELLUM
PONS
MEDULLA
SPINAL CORD
RHOMBEN-CEPHALON
Olfactory bulb

Figure 21. The lateral view of a 3-D model of C966's brain and upper cervical spinal cord (part of the Carnegie Collection at the National Museum of Health and Medicine) shows the exact locations and cutting angles of the illustrated sections of C966 in the following pages. The small inset identifies the major structural features. The medullary velum was not reconstructed so that the rhombencephalic superventricle appears as an open gap beneath the cerebellum.

PLATE 16A

CR 23 mm, GW 8.4, C966
Frontal/horizontal
Level 1: Section 10

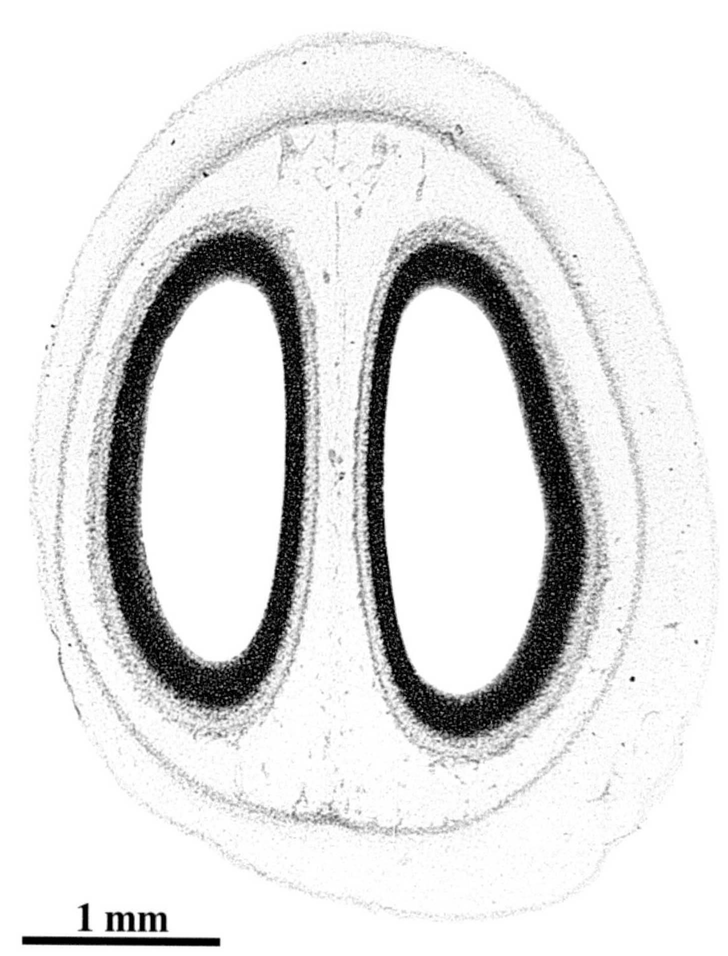
1 mm

PLATE 16B

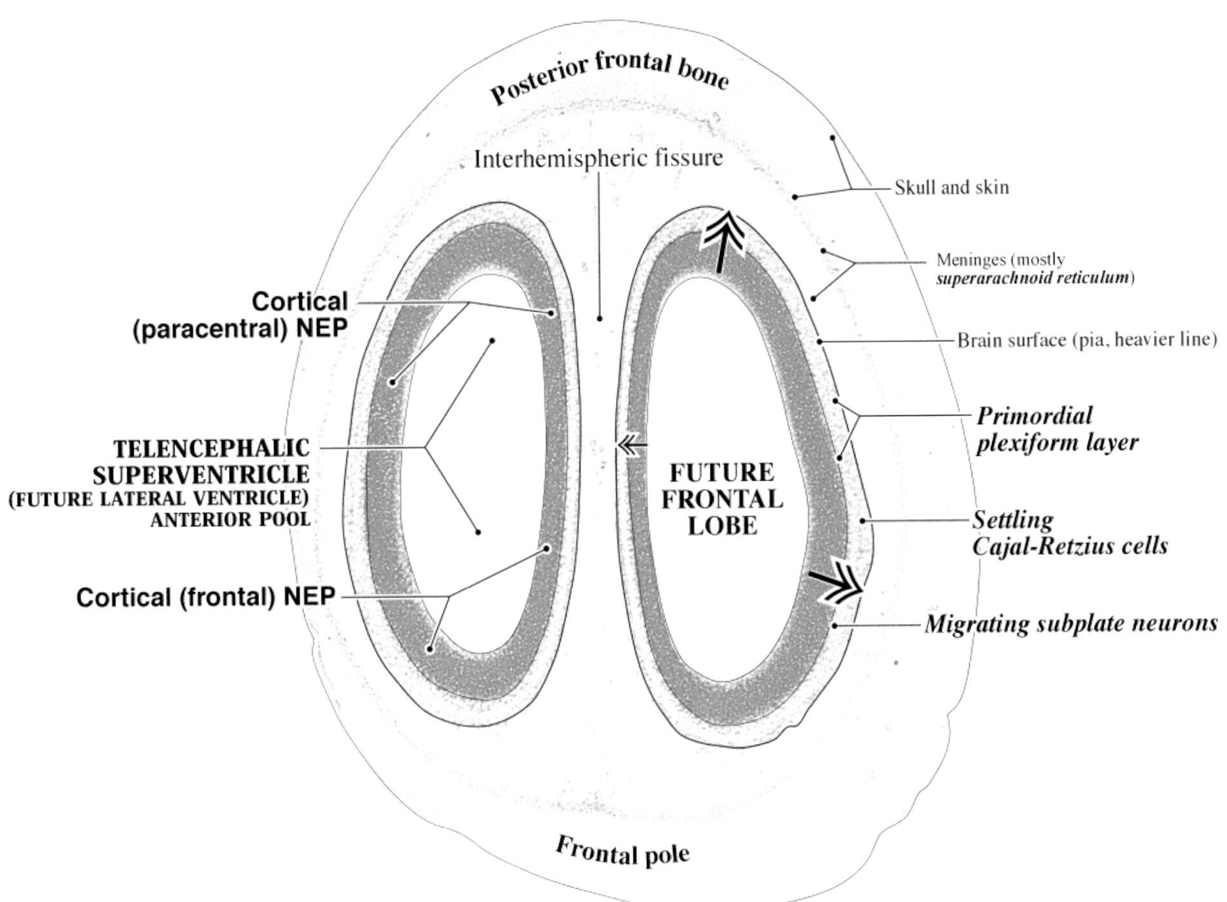
Posterior frontal bone
Interhemispheric fissure
Skull and skin
Meninges (mostly
superarachnoid reticulum)
Brain surface (pia, heavier line)
Primordial
plexiform layer
Settling
Cajal-Retzius cells
Migrating subplate neurons
Cortical
(paracentral) NEP
TELENCEPHALIC
SUPERVENTRICLE
(FUTURE LATERAL VENTRICLE)
ANTERIOR POOL
Cortical (frontal) NEP
FUTURE
FRONTAL
LOBE
Frontal pole

NEP - Neuroepithelium

FONT KEY:
VENTRICULAR DIVISIONS - CAPITALS
Germinal zone - Helvetica bold
Transient structure - Times bold italic
Permanent structure - Times Roman or **Bold**

Arrows indicate the presumed ***direction of neuron migration*** from neuroepithelial sources.

PLATE 17A

CR 23 mm, GW 8.4, C966
Frontal/horizontal
Level 2: Section 34

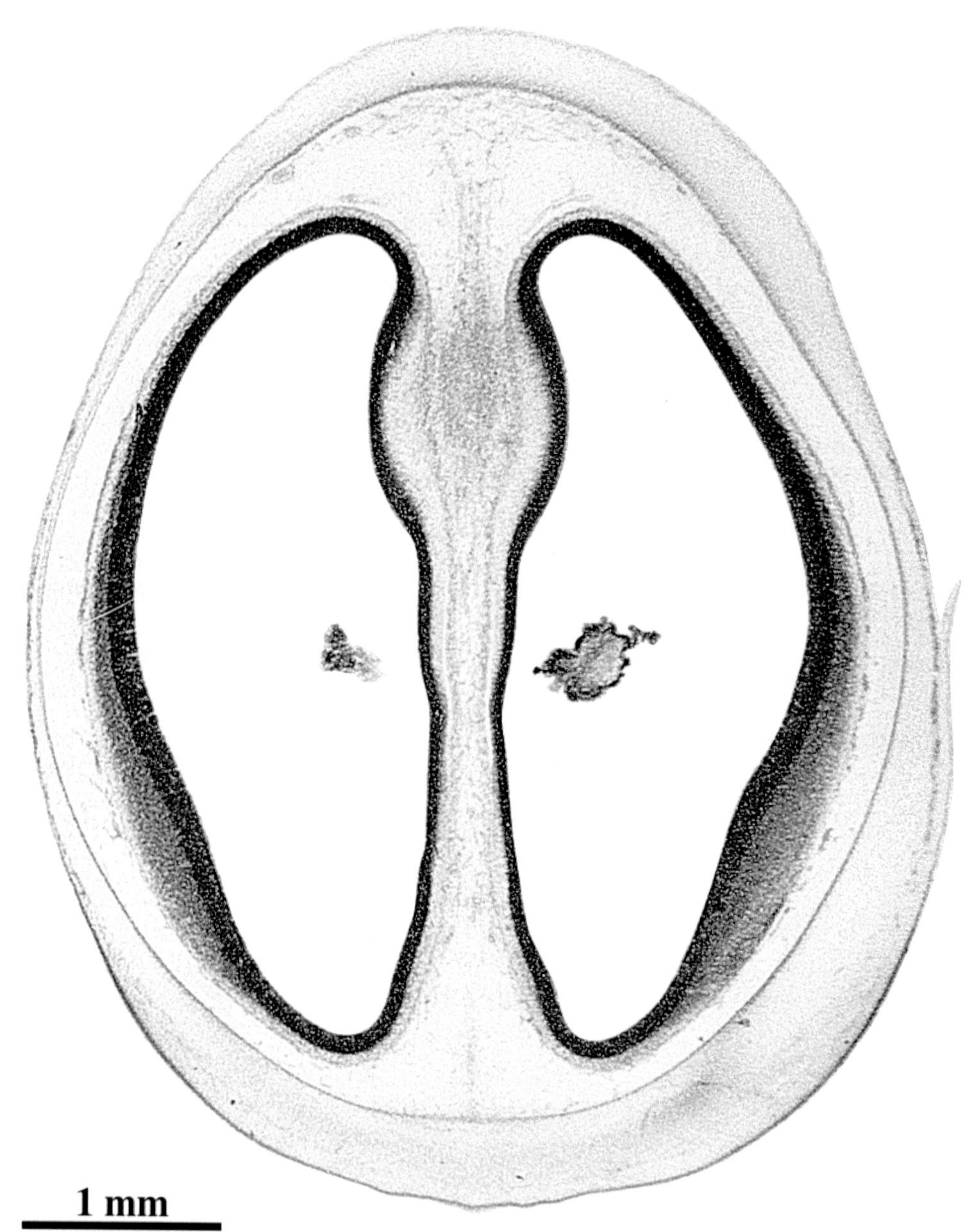
1 mm

PLATE 17B

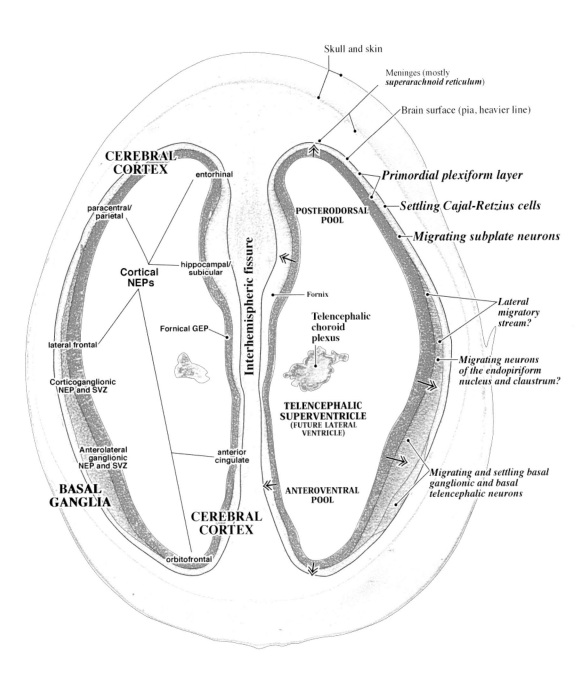

Skull and skin

Meninges (mostly ***superarachnoid reticulum***)

Brain surface (pia, heavier line)

Primordial plexiform layer

Settling Cajal-Retzius cells

Migrating subplate neurons

CEREBRAL CORTEX

entorhinal

paracentral/ parietal

POSTERODORSAL POOL

hippocampal/ subicular

Cortical NEPs

Interhemispheric fissure

Fornix

Lateral migratory stream?

Telencephalic choroid plexus

Fornical GEP

lateral frontal

Migrating neurons of the endopiriform nucleus and claustrum?

Corticoganglionic NEP and SVZ

TELENCEPHALIC SUPERVENTRICLE
(FUTURE LATERAL VENTRICLE)

Anterolateral ganglionic NEP and SVZ

anterior cingulate

Migrating and settling basal ganglionic and basal telencephalic neurons

BASAL GANGLIA

ANTEROVENTRAL POOL

CEREBRAL CORTEX

orbitofrontal

ABBREVIATIONS:
GEP - Glioepithelium
NEP - Neuroepithelium
SVZ - Subventricular zone

FONT KEY:
VENTRICULAR DIVISIONS - CAPITALS
Germinal zone - Helvetica bold
Transient structure - Times bold italic
Permanent structure - Times Roman or **Bold**

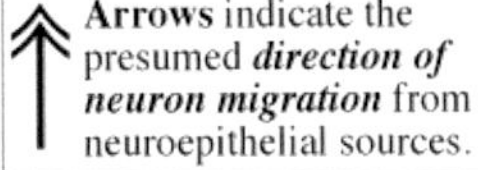

Arrows indicate the presumed ***direction of neuron migration*** from neuroepithelial sources.

PLATE 18A

CR 23 mm, GW 8.4, C966
Frontal/horizontal
Level 3: Section 40

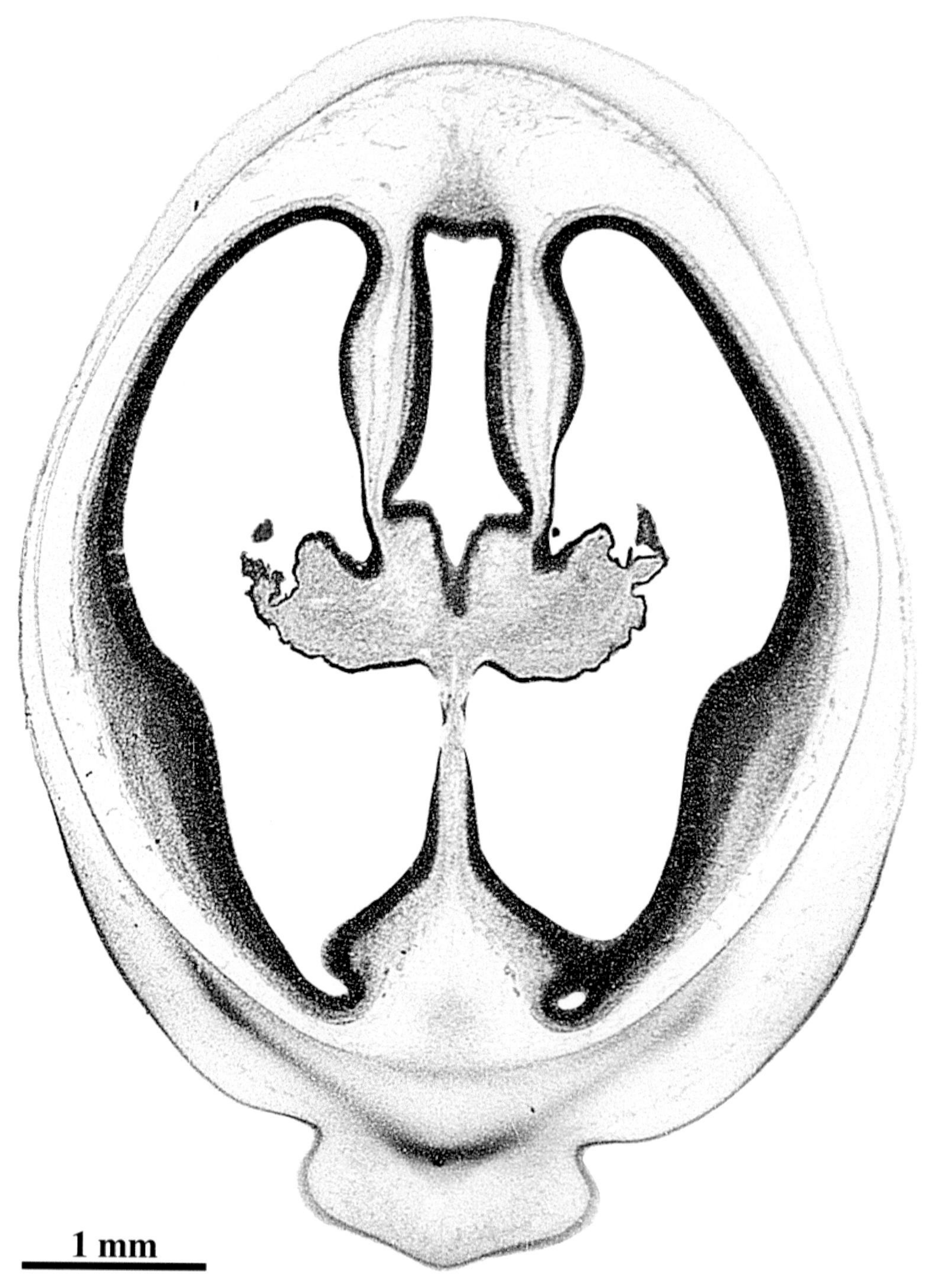
1 mm

PLATE 18B

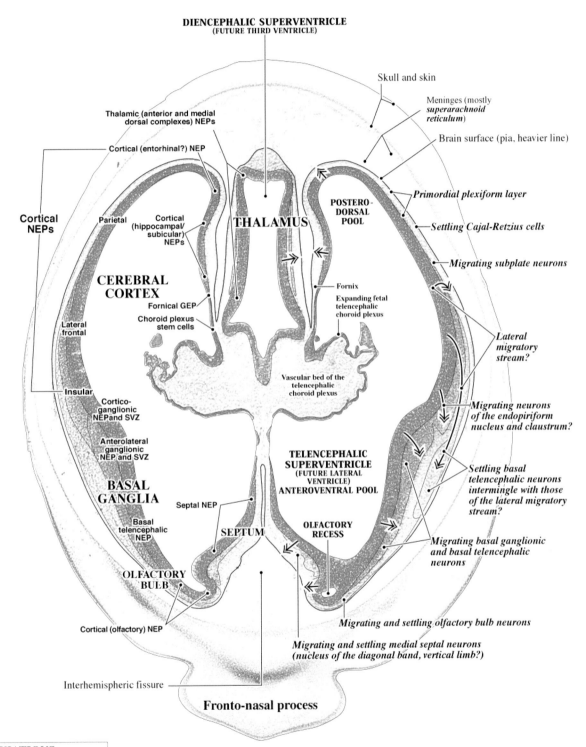

DIENCEPHALIC SUPERVENTRICLE
(FUTURE THIRD VENTRICLE)

Skull and skin

Meninges (mostly ***superarachnoid reticulum***)

Brain surface (pia, heavier line)

Primordial plexiform layer

Settling Cajal-Retzius cells

Migrating subplate neurons

Lateral migratory stream?

Migrating neurons of the endopiriform nucleus and claustrum?

Settling basal telencephalic neurons intermingle with those of the lateral migratory stream?

Migrating basal ganglionic and basal telencephalic neurons

Migrating and settling olfactory bulb neurons

Migrating and settling medial septal neurons (nucleus of the diagonal band, vertical limb?)

Thalamic (anterior and medial dorsal complexes) NEPs

Cortical (entorhinal?) NEP

Cortical NEPs

Parietal

Cortical (hippocampal/ subicular) NEPs

THALAMUS

POSTERO-DORSAL POOL

CEREBRAL CORTEX

Fornical GEP

Choroid plexus stem cells

Lateral frontal

Fornix

Expanding fetal telencephalic choroid plexus

Vascular bed of the telencephalic choroid plexus

Insular

Cortico-ganglionic NEP and SVZ

Anterolateral ganglionic NEP and SVZ

TELENCEPHALIC SUPERVENTRICLE
(FUTURE LATERAL VENTRICLE)
ANTEROVENTRAL POOL

BASAL GANGLIA

Septal NEP

Basal telencephalic NEP

SEPTUM

OLFACTORY RECESS

OLFACTORY BULB

Cortical (olfactory) NEP

Interhemispheric fissure

Fronto-nasal process

ABBREVIATIONS:
GEP - Glioepithelium
NEP - Neuroepithelium
SVZ - Subventricular zone

FONT KEY:
VENTRICULAR DIVISIONS - CAPITALS
Germinal zone - Helvetica bold
Transient structure - Times bold italic
Permanent structure - Times Roman or **Bold**

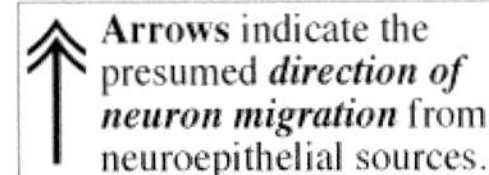

Arrows indicate the presumed ***direction of neuron migration*** from neuroepithelial sources.

PLATE 19A

CR 23 mm, GW 8.4, C966
Frontal/horizontal
Level 4: Section 52

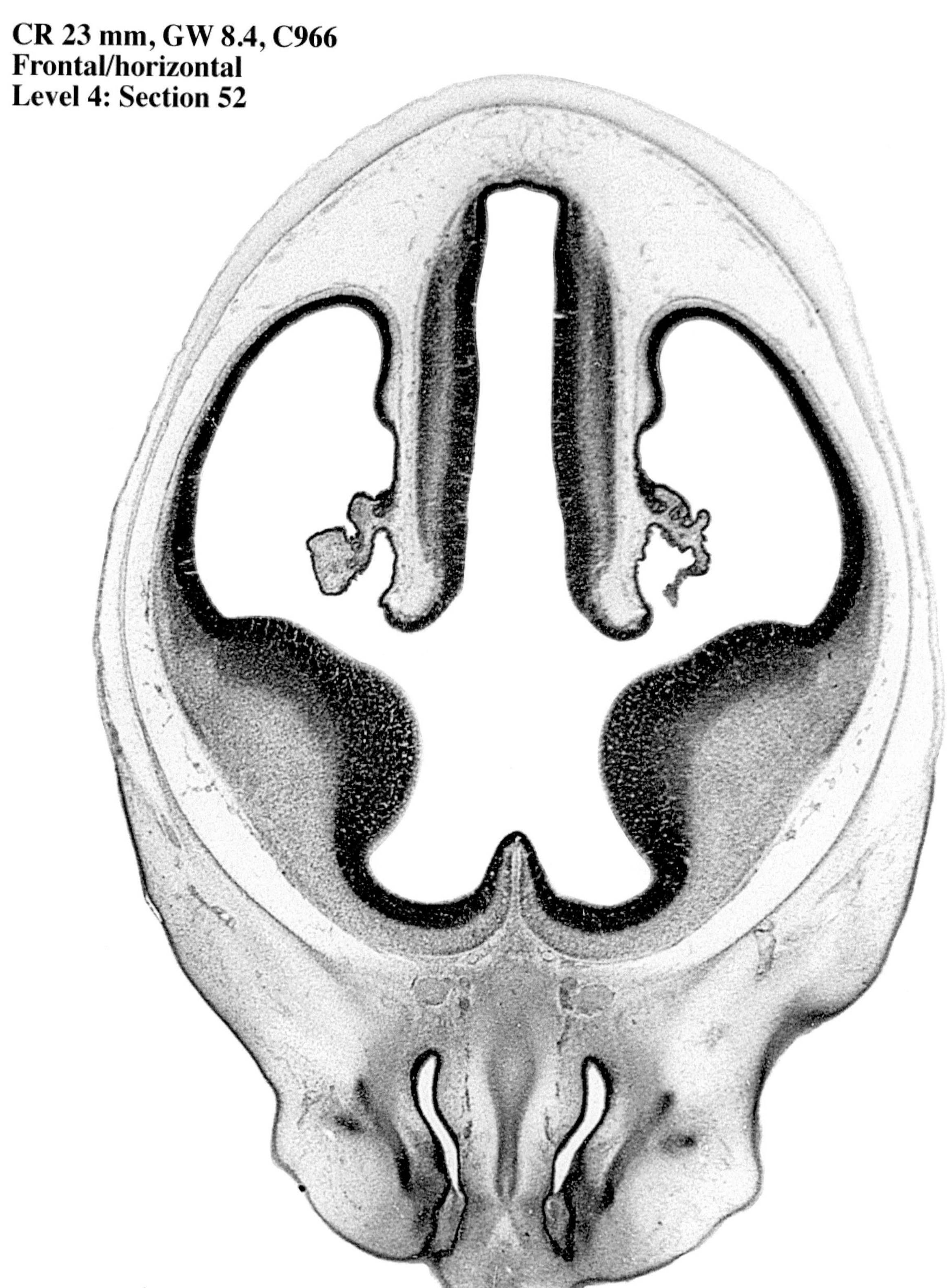

1 mm

PLATE 19B

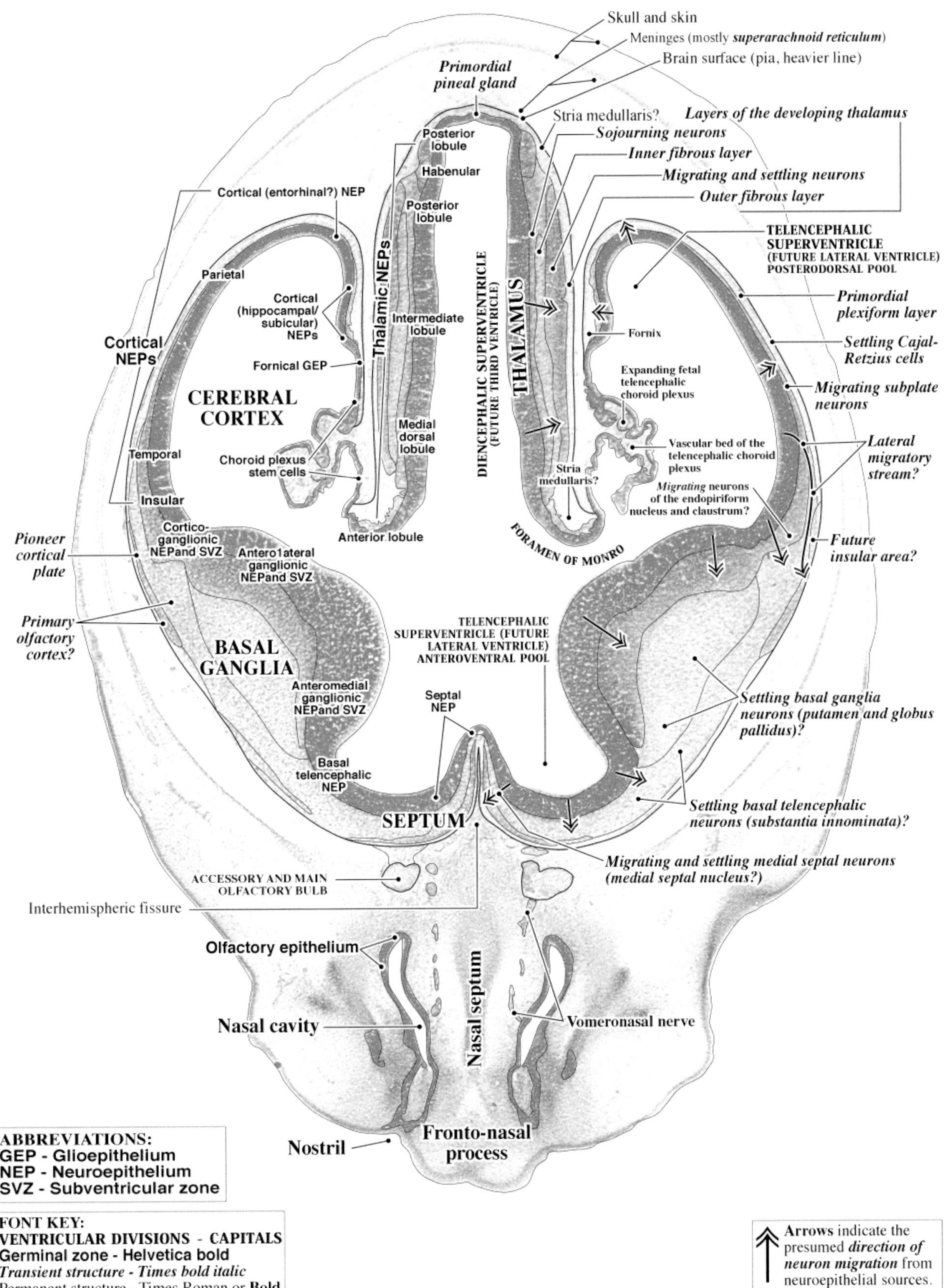
Skull and skin
Meninges (mostly *superarachnoid reticulum*)
Brain surface (pia, heavier line)
Primordial pineal gland
Stria medullaris?
Layers of the developing thalamus
Sojourning neurons
Inner fibrous layer
Migrating and settling neurons
Outer fibrous layer
Posterior lobule
Habenular
Posterior lobule
Cortical (entorhinal?) NEP
Parietal
Cortical (hippocampal/ subicular) NEPs
Thalamic NEPs
Intermediate lobule
DIENCEPHALIC SUPERVENTRICLE (FUTURE THIRD VENTRICLE)
THALAMUS
TELENCEPHALIC SUPERVENTRICLE (FUTURE LATERAL VENTRICLE) POSTERODORSAL POOL
Primordial plexiform layer
Fornix
Settling Cajal-Retzius cells
Cortical NEPs
Fornical GEP
Expanding fetal telencephalic choroid plexus
Migrating subplate neurons
CEREBRAL CORTEX
Medial dorsal lobule
Vascular bed of the telencephalic choroid plexus
Lateral migratory stream?
Temporal
Choroid plexus stem cells
Stria medullaris?
Migrating neurons of the endopiriform nucleus and claustrum?
Insular
Cortico-ganglionic NEPand SVZ
Anterior lobule
FORAMEN OF MONRO
Pioneer cortical plate
Antero1ateral ganglionic NEPand SVZ
Future insular area?
Primary olfactory cortex?
BASAL GANGLIA
TELENCEPHALIC SUPERVENTRICLE (FUTURE LATERAL VENTRICLE) ANTEROVENTRAL POOL
Anteromedial ganglionic NEPand SVZ
Septal NEP
Settling basal ganglia neurons (putamen and globus pallidus)?
Basal telencephalic NEP
Settling basal telencephalic neurons (substantia innominata)?
SEPTUM
Migrating and settling medial septal neurons (medial septal nucleus?)
ACCESSORY AND MAIN OLFACTORY BULB
Interhemispheric fissure
Olfactory epithelium
Nasal septum
Nasal cavity
Vomeronasal nerve
Nostril
Fronto-nasal process
ABBREVIATIONS:
GEP - Glioepithelium
NEP - Neuroepithelium
SVZ - Subventricular zone
FONT KEY:
VENTRICULAR DIVISIONS - CAPITALS
Germinal zone - Helvetica bold
Transient structure - Times bold italic
Permanent structure - Times Roman or **Bold**
Arrows indicate the presumed ***direction of neuron migration*** from neuroepithelial sources.

PLATE 20A

CR 23 mm, GW 8.4, C966
Frontal/horizontal
Level 5: Section 58

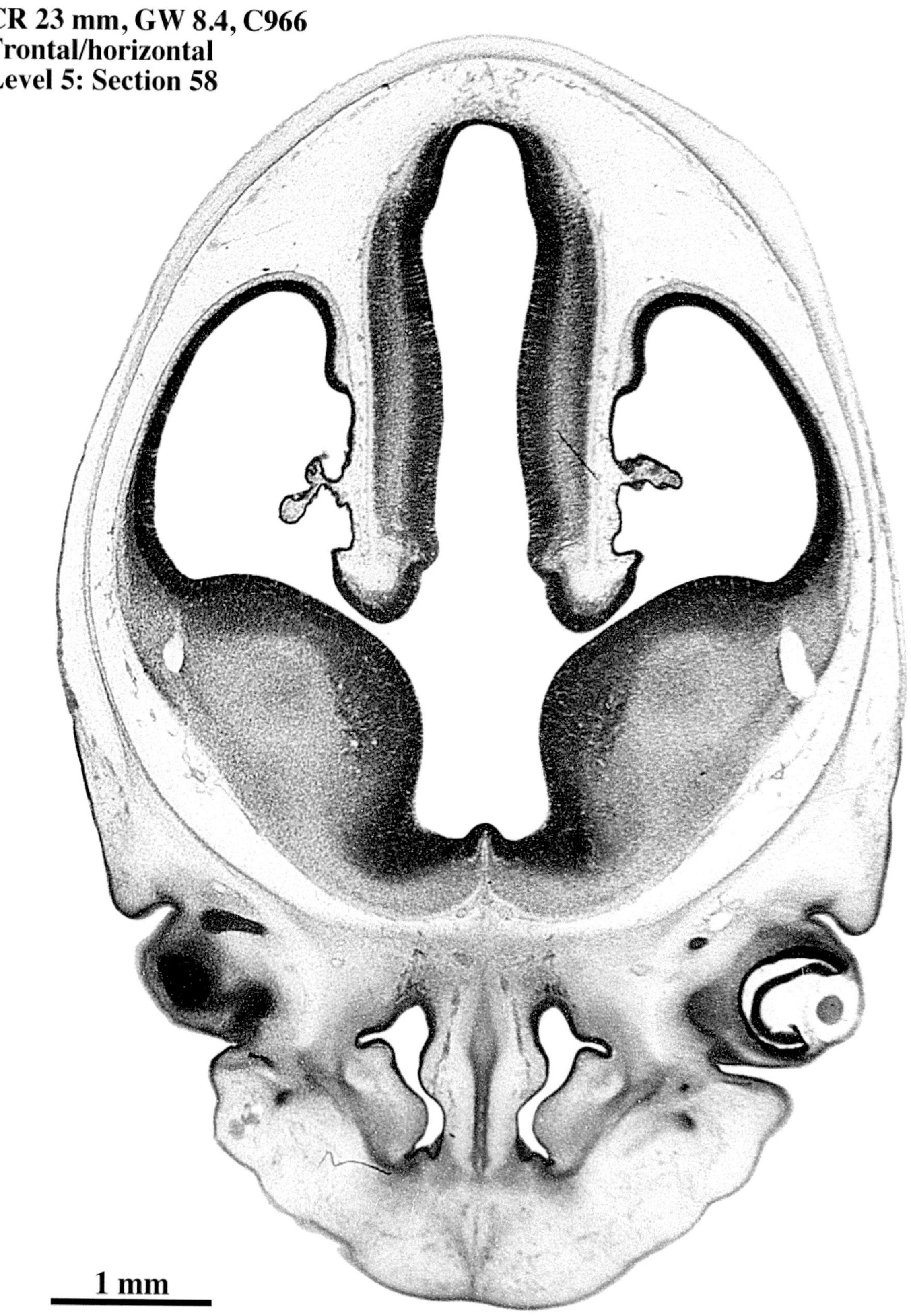
1 mm

PLATE 20B

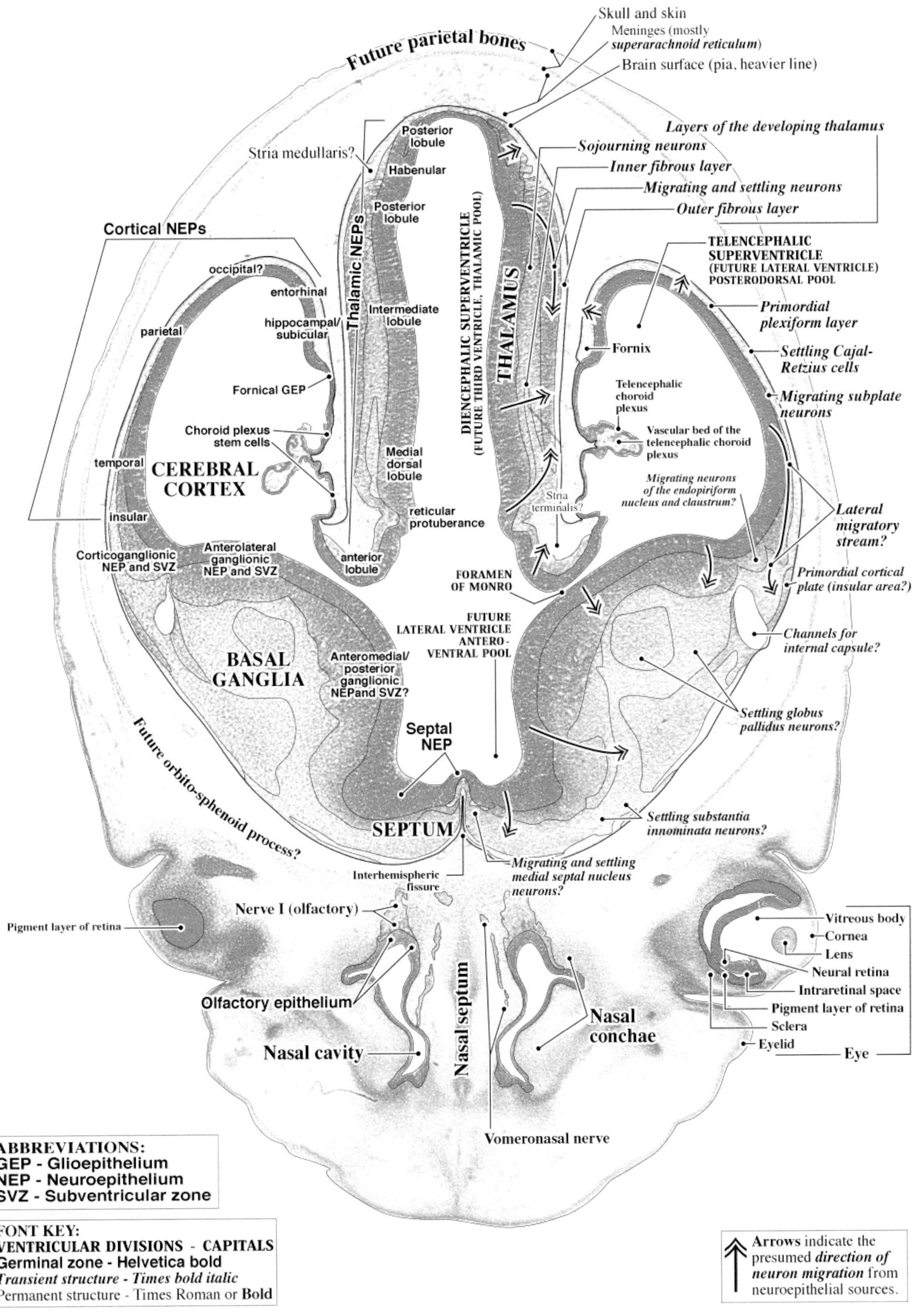

Skull and skin
Meninges (mostly superarachnoid reticulum)
Brain surface (pia, heavier line)
Future parietal bones
Layers of the developing thalamus
Sojourning neurons
Inner fibrous layer
Migrating and settling neurons
Outer fibrous layer
Posterior lobule
Stria medullaris?
Habenular
Posterior lobule
Cortical NEPs
Thalamic NEPs
DIENCEPHALIC SUPERVENTRICLE (FUTURE THIRD VENTRICLE, THALAMIC POOL)
THALAMUS
TELENCEPHALIC SUPERVENTRICLE (FUTURE LATERAL VENTRICLE) POSTERODORSAL POOL
occipital?
entorhinal
hippocampal/ subicular
parietal
Intermediate lobule
Primordial plexiform layer
Fornix
Settling Cajal-Retzius cells
Fornical GEP
Telencephalic choroid plexus
Migrating subplate neurons
Choroid plexus stem cells
Vascular bed of the telencephalic choroid plexus
temporal
CEREBRAL CORTEX
Medial dorsal lobule
Migrating neurons of the endopiriform nucleus and claustrum?
Stria terminalis?
insular
reticular protuberance
Lateral migratory stream?
Corticoganglionic NEP and SVZ
Anterolateral ganglionic NEP and SVZ
anterior lobule
FORAMEN OF MONRO
Primordial cortical plate (insular area?)
FUTURE LATERAL VENTRICLE ANTERO-VENTRAL POOL
Channels for internal capsule?
BASAL GANGLIA
Anteromedial/ posterior ganglionic NEPand SVZ?
Settling globus pallidus neurons?
Septal NEP
Future orbito-sphenoid process?
SEPTUM
Settling substantia innominata neurons?
Migrating and settling medial septal nucleus neurons?
Interhemispheric fissure
Nerve I (olfactory)
Pigment layer of retina
Vitreous body
Cornea
Lens
Neural retina
Intraretinal space
Pigment layer of retina
Sclera
Eyelid
Eye
Olfactory epithelium
Nasal septum
Nasal conchae
Nasal cavity
Vomeronasal nerve

ABBREVIATIONS:
GEP - Glioepithelium
NEP - Neuroepithelium
SVZ - Subventricular zone

FONT KEY:
VENTRICULAR DIVISIONS - CAPITALS
Germinal zone - Helvetica bold
Transient structure - Times bold italic
Permanent structure - Times Roman or **Bold**

Arrows indicate the presumed ***direction of neuron migration*** from neuroepithelial sources.

PLATE 21A

CR 23 mm, GW 8.4, C966
Frontal/horizontal
Level 6: Section 70

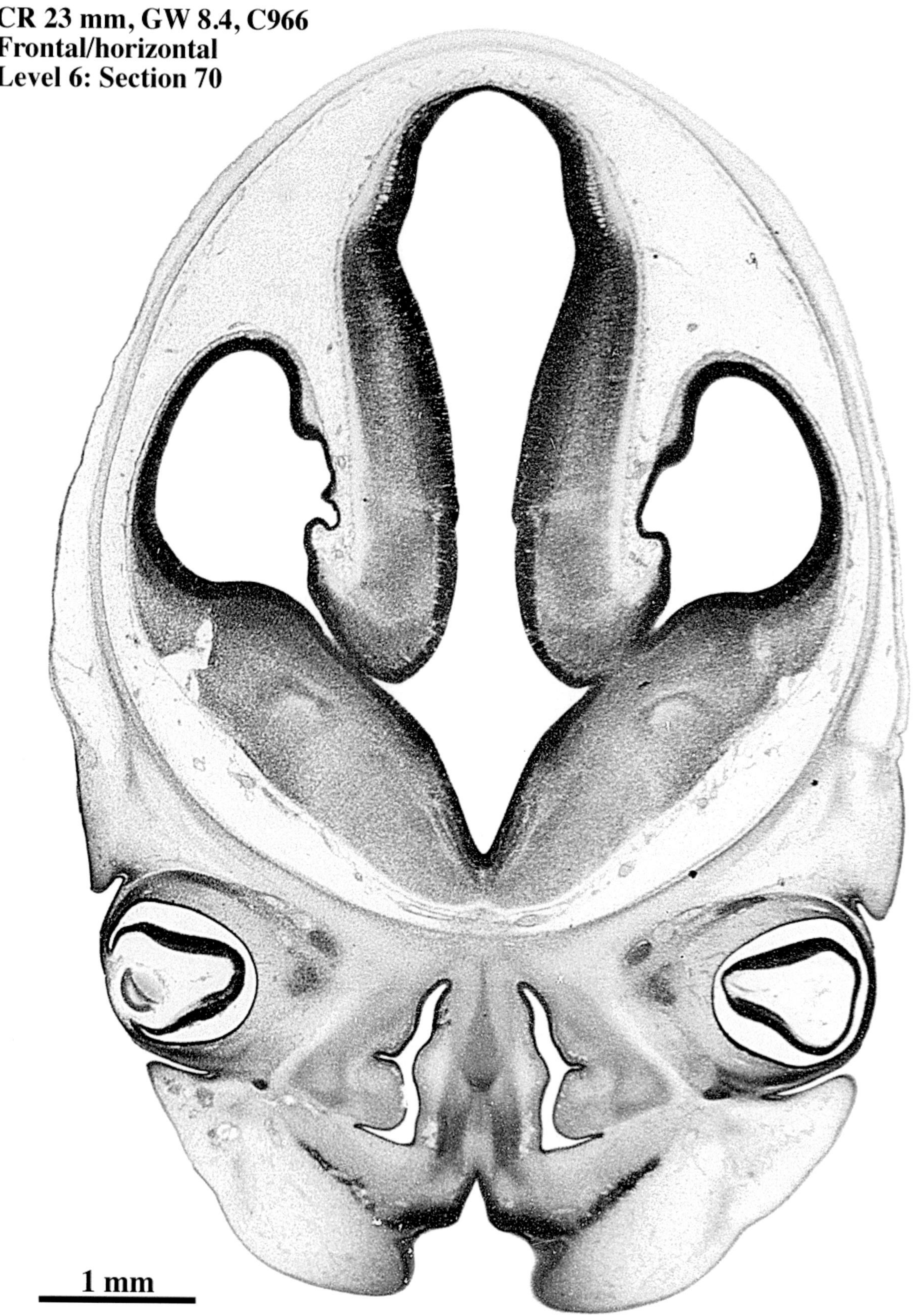
1 mm

PLATE 21B

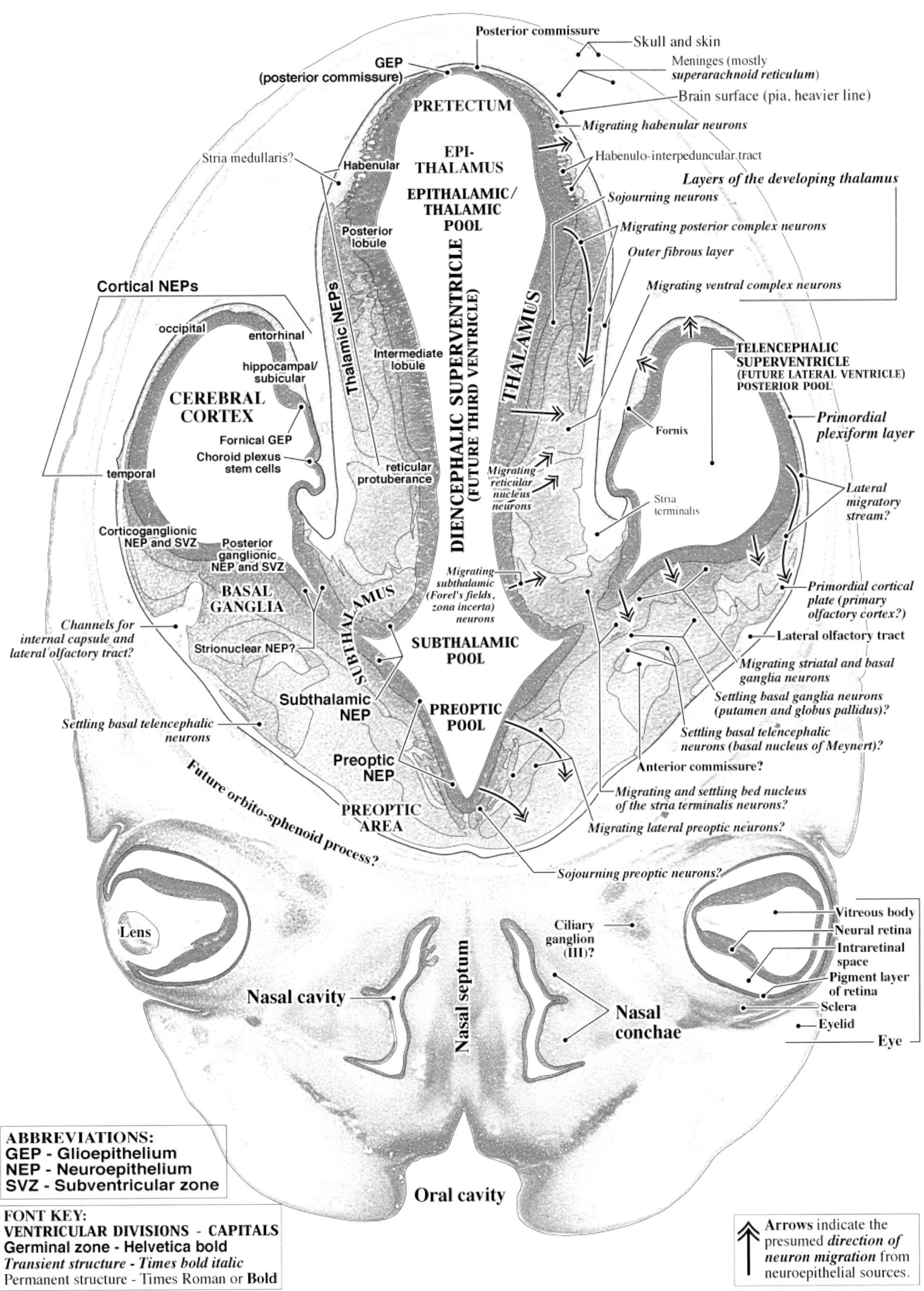
Posterior commissure
Skull and skin
GEP
(posterior commissure)
Meninges (mostly
superarachnoid reticulum)
Brain surface (pia, heavier line)
PRETECTUM
Migrating habenular neurons
EPI-
THALAMUS
Stria medullaris?
Habenular
Habenulo-interpeduncular tract
Layers of the developing thalamus
EPITHALAMIC/
THALAMIC
POOL
Sojourning neurons
Posterior
lobule
Migrating posterior complex neurons
Outer fibrous layer
Cortical NEPs
Migrating ventral complex neurons
Thalamic NEPs
DIENCEPHALIC SUPERVENTRICLE
(FUTURE THIRD VENTRICLE)
THALAMUS
occipital
entorhinal
Intermediate
lobule
TELENCEPHALIC
SUPERVENTRICLE
(FUTURE LATERAL VENTRICLE)
POSTERIOR POOL
hippocampal/
subicular
CEREBRAL
CORTEX
Fornix
Primordial
plexiform layer
Fornical GEP
Choroid plexus
stem cells
temporal
reticular
protuberance
Migrating
reticular
nucleus
neurons
Stria
terminalis
Lateral
migratory
stream?
Corticoganglionic
NEP and SVZ
Posterior
ganglionic
NEP and SVZ
Migrating
subthalamic
(Forel's fields,
zona incerta)
neurons
BASAL
GANGLIA
Primordial cortical
plate (primary
olfactory cortex?)
Channels for
internal capsule and
lateral olfactory tract?
Strionuclear NEP?
SUBTHALAMUS
SUBTHALAMIC
POOL
Lateral olfactory tract
Migrating striatal and basal
ganglia neurons
Subthalamic
NEP
PREOPTIC
POOL
Settling basal ganglia neurons
(putamen and globus pallidus)?
Settling basal telencephalic
neurons
Settling basal telencephalic
neurons (basal nucleus of Meynert)?
Preoptic
NEP
Anterior commissure?
Future orbito-sphenoid process?
PREOPTIC
AREA
Migrating and settling bed nucleus
of the stria terminalis neurons?
Migrating lateral preoptic neurons?
Sojourning preoptic neurons?
Lens
Ciliary
ganglion
(III)?
Vitreous body
Neural retina
Intraretinal
space
Pigment layer
of retina
Sclera
Eyelid
Eye
Nasal cavity
Nasal septum
Nasal
conchae
Oral cavity
ABBREVIATIONS:
GEP - Glioepithelium
NEP - Neuroepithelium
SVZ - Subventricular zone
FONT KEY:
VENTRICULAR DIVISIONS - CAPITALS
Germinal zone - Helvetica bold
Transient structure - Times bold italic
Permanent structure - Times Roman or Bold
Arrows indicate the presumed direction of neuron migration from neuroepithelial sources.

PLATE 22A

CR 23 mm, GW 8.4, C966
Frontal/horizontal
Level 7: Section 76

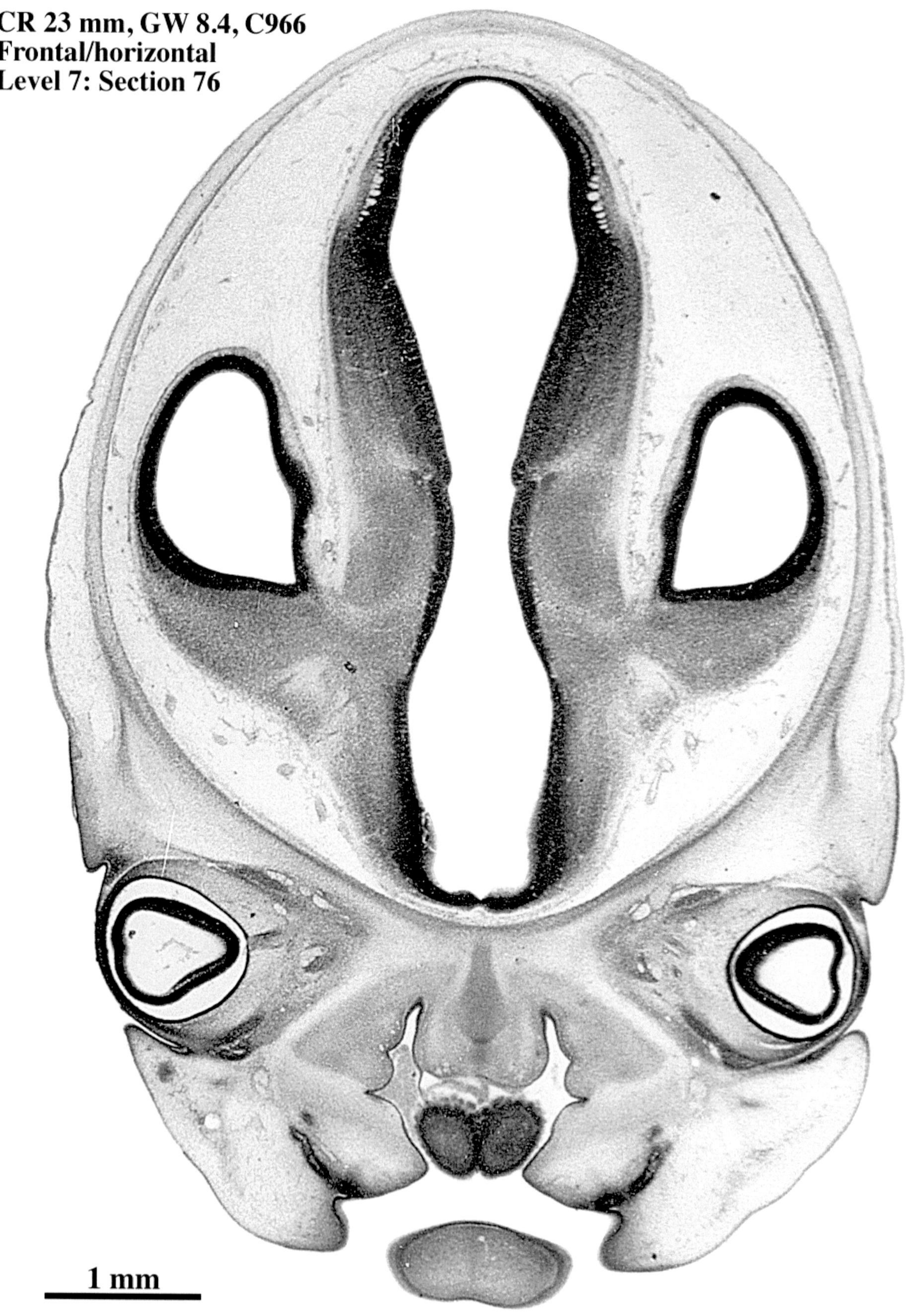
1 mm

PLATE 22B

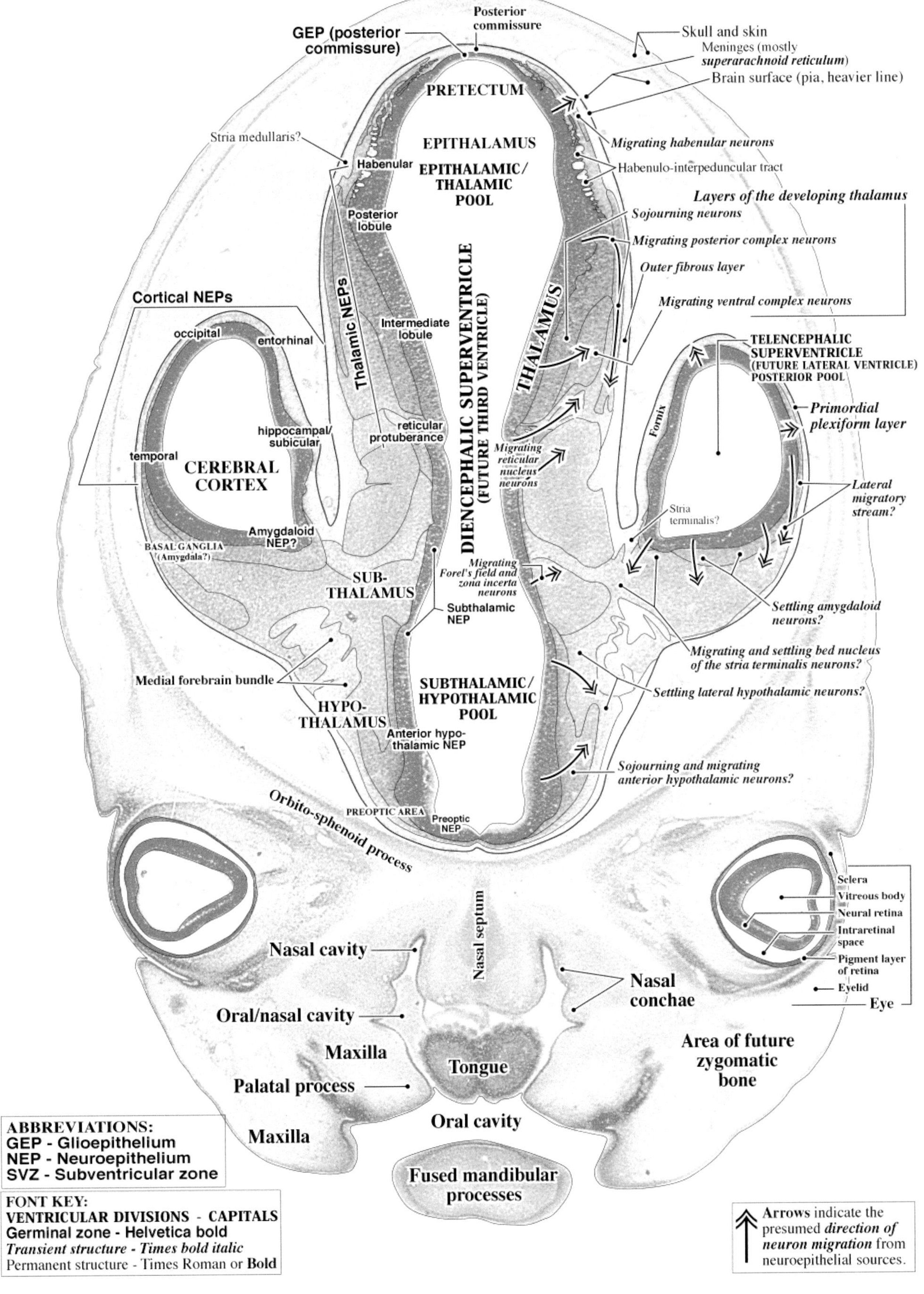

Posterior commissure
GEP (posterior commissure)
Skull and skin
Meninges (mostly superarachnoid reticulum)
Brain surface (pia, heavier line)
PRETECTUM
Stria medullaris?
EPITHALAMUS
Migrating habenular neurons
Habenular
EPITHALAMIC/ THALAMIC POOL
Habenulo-interpeduncular tract
Layers of the developing thalamus
Sojourning neurons
Posterior lobule
Migrating posterior complex neurons
Outer fibrous layer
Cortical NEPs
Thalamic NEPs
THALAMUS
Migrating ventral complex neurons
DIENCEPHALIC SUPERVENTRICLE (FUTURE THIRD VENTRICLE)
Intermediate lobule
occipital
entorhinal
TELENCEPHALIC SUPERVENTRICLE (FUTURE LATERAL VENTRICLE) POSTERIOR POOL
Fornix
Primordial plexiform layer
hippocampal/ subicular
reticular protuberance
Migrating reticular nucleus neurons
temporal
CEREBRAL CORTEX
Lateral migratory stream?
Stria terminalis?
Amygdaloid NEP?
BASAL GANGLIA (Amygdala?)
Migrating Forel's field and zona incerta neurons
SUB-THALAMUS
Settling amygdaloid neurons?
Subthalamic NEP
Migrating and settling bed nucleus of the stria terminalis neurons?
Medial forebrain bundle
SUBTHALAMIC/ HYPOTHALAMIC POOL
Settling lateral hypothalamic neurons?
HYPO-THALAMUS
Anterior hypo-thalamic NEP
Sojourning and migrating anterior hypothalamic neurons?
Orbito-sphenoid process
PREOPTIC AREA
Preoptic NEP
Nasal septum
Sclera
Vitreous body
Neural retina
Intraretinal space
Pigment layer of retina
Eyelid
Eye
Nasal cavity
Nasal conchae
Oral/nasal cavity
Maxilla
Tongue
Area of future zygomatic bone
Palatal process
Oral cavity
Maxilla
Fused mandibular processes

ABBREVIATIONS:
GEP - Glioepithelium
NEP - Neuroepithelium
SVZ - Subventricular zone

FONT KEY:
VENTRICULAR DIVISIONS - CAPITALS
Germinal zone - Helvetica bold
Transient structure - Times bold italic
Permanent structure - Times Roman or **Bold**

Arrows indicate the presumed ***direction of neuron migration*** from neuroepithelial sources.

PLATE 23A

CR 23 mm, GW 8.4, C966
Frontal/horizontal
Level 8: Section 84

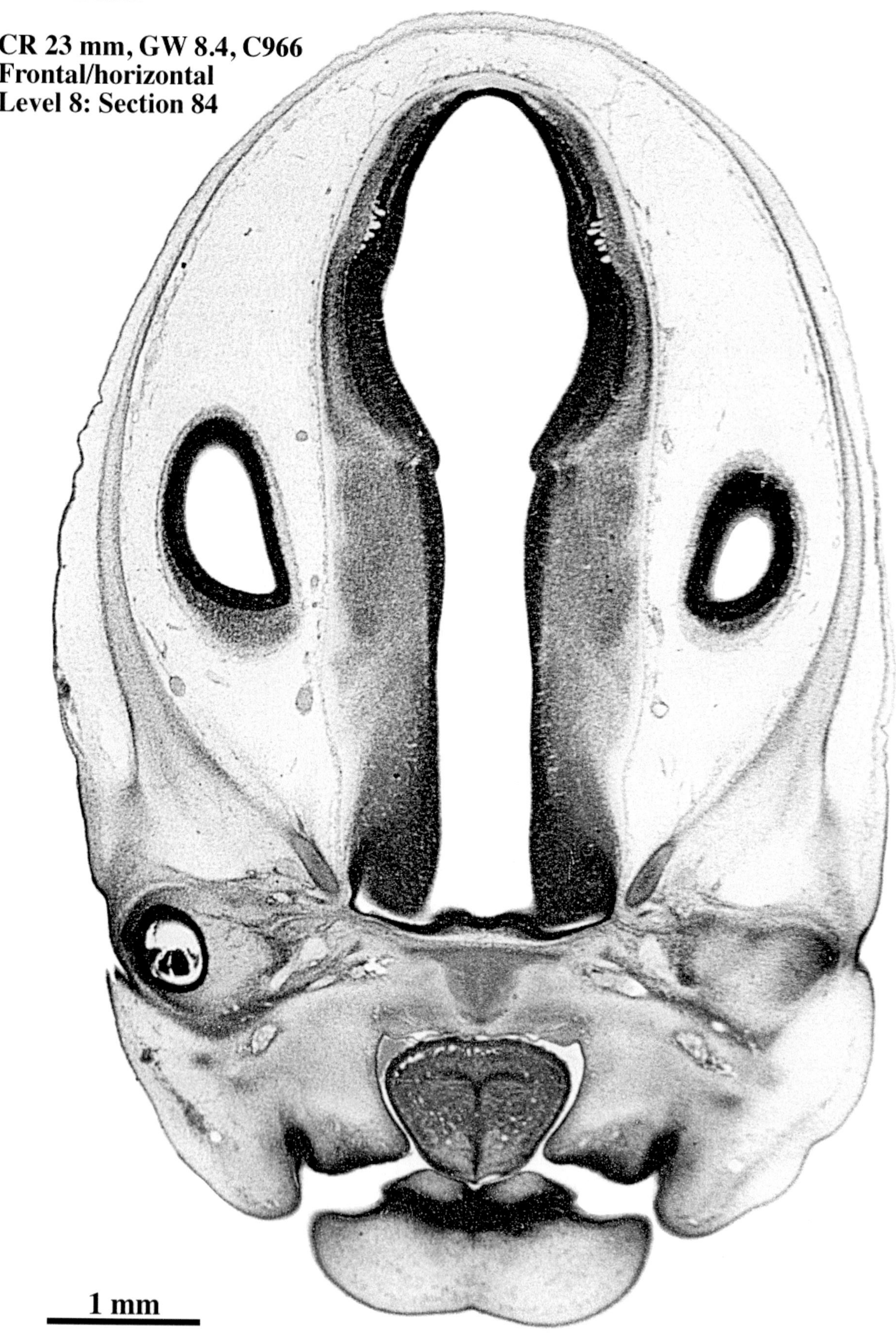

1 mm

PLATE 23B

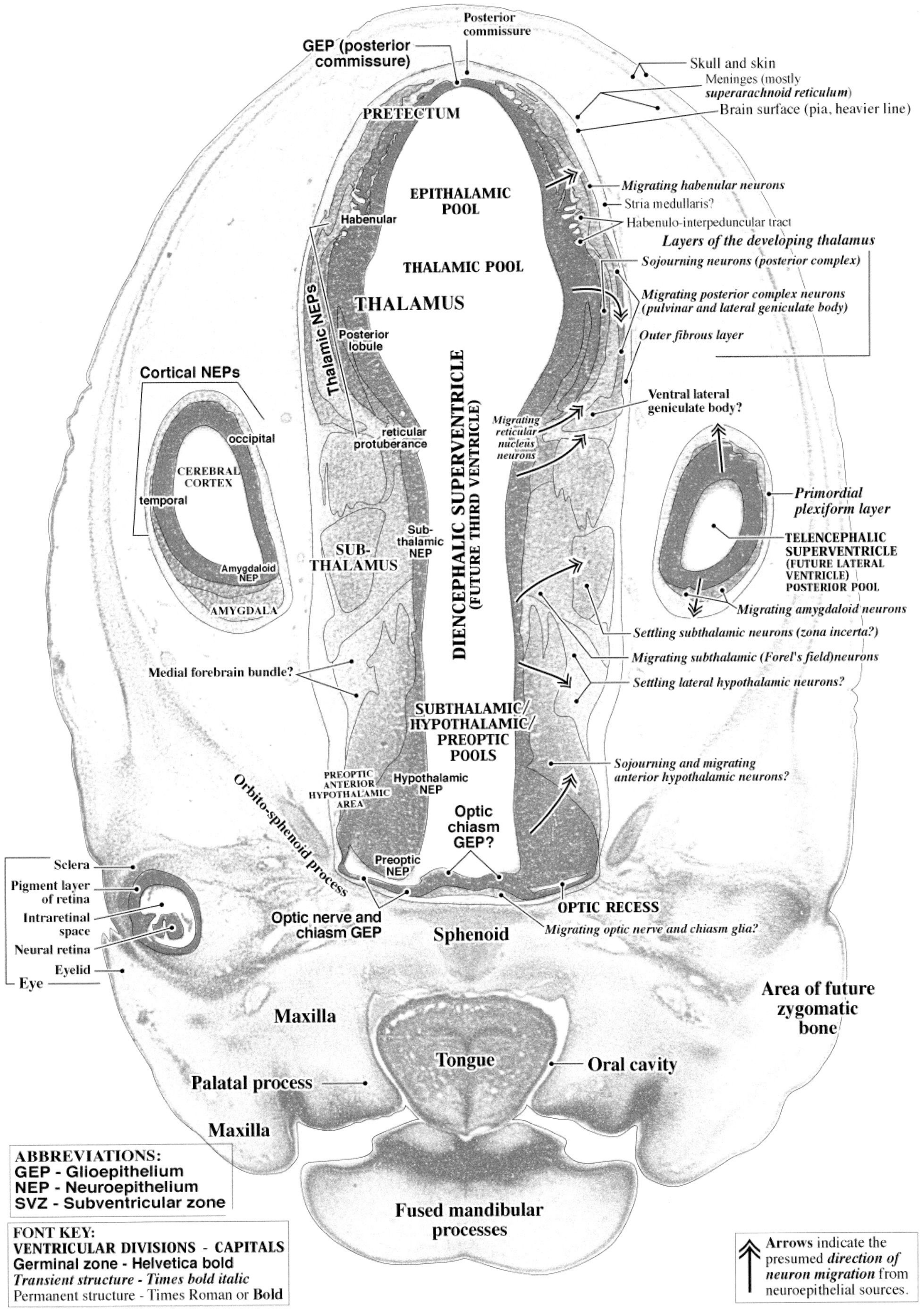

Posterior commissure
GEP (posterior commissure)
Skull and skin
Meninges (mostly *superarachnoid reticulum*)
Brain surface (pia, heavier line)
PRETECTUM
EPITHALAMIC POOL
Habenular
Migrating habenular neurons
Stria medullaris?
Habenulo-interpeduncular tract
Layers of the developing thalamus
THALAMIC POOL
Sojourning neurons (posterior complex)
THALAMUS
Thalamic NEPs
Migrating posterior complex neurons (pulvinar and lateral geniculate body)
Posterior lobule
Outer fibrous layer
Cortical NEPs
DIENCEPHALIC SUPERVENTRICLE (FUTURE THIRD VENTRICLE)
Ventral lateral geniculate body?
occipital
reticular protuberance
Migrating reticular nucleus neurons
CEREBRAL CORTEX
temporal
Primordial plexiform layer
TELENCEPHALIC SUPERVENTRICLE (FUTURE LATERAL VENTRICLE) POSTERIOR POOL
Sub-thalamic NEP
SUB-THALAMUS
Amygdaloid NEP
AMYGDALA
Migrating amygdaloid neurons
Settling subthalamic neurons (zona incerta?)
Migrating subthalamic (Forel's field)neurons
Medial forebrain bundle?
Settling lateral hypothalamic neurons?
SUBTHALAMIC/ HYPOTHALAMIC/ PREOPTIC POOLS
Sojourning and migrating anterior hypothalamic neurons?
PREOPTIC ANTERIOR HYPOTHALAMIC AREA
Hypothalamic NEP
Orbito-sphenoid process
Optic chiasm GEP?
Sclera
Pigment layer of retina
Intraretinal space
Neural retina
Eyelid
Eye
Preoptic NEP
OPTIC RECESS
Optic nerve and chiasm GEP
Sphenoid
Migrating optic nerve and chiasm glia?
Maxilla
Area of future zygomatic bone
Tongue
Oral cavity
Palatal process
Maxilla
ABBREVIATIONS:
GEP - Glioepithelium
NEP - Neuroepithelium
SVZ - Subventricular zone
Fused mandibular processes
FONT KEY:
VENTRICULAR DIVISIONS - CAPITALS
Germinal zone - Helvetica bold
Transient structure - Times bold italic
Permanent structure - Times Roman or **Bold**
Arrows indicate the presumed *direction of neuron migration* from neuroepithelial sources.

PLATE 24A

CR 23 mm, GW 8.4, C966
Frontal/horizontal
Level 9: Section 92

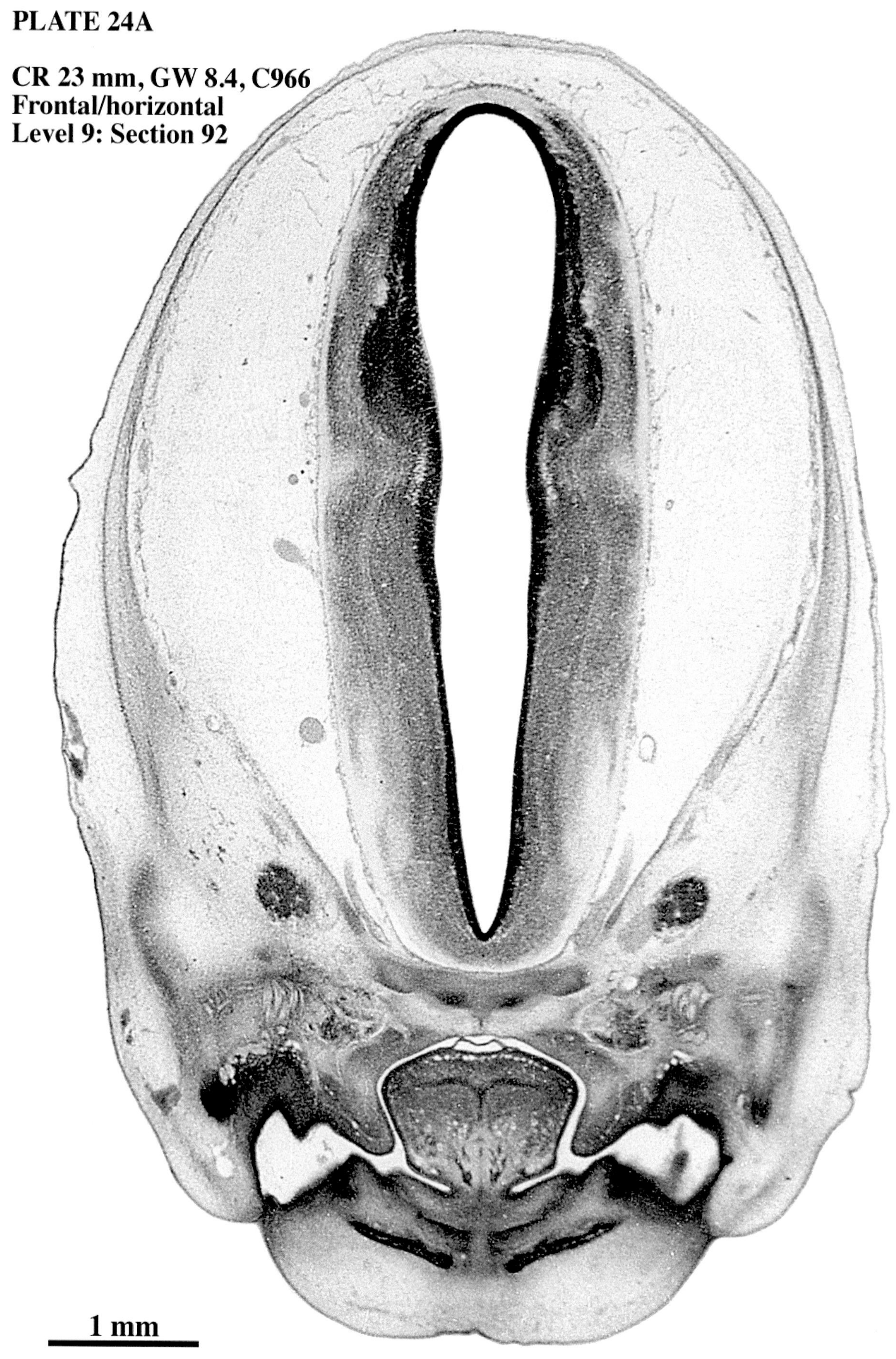
1 mm

PLATE 24B

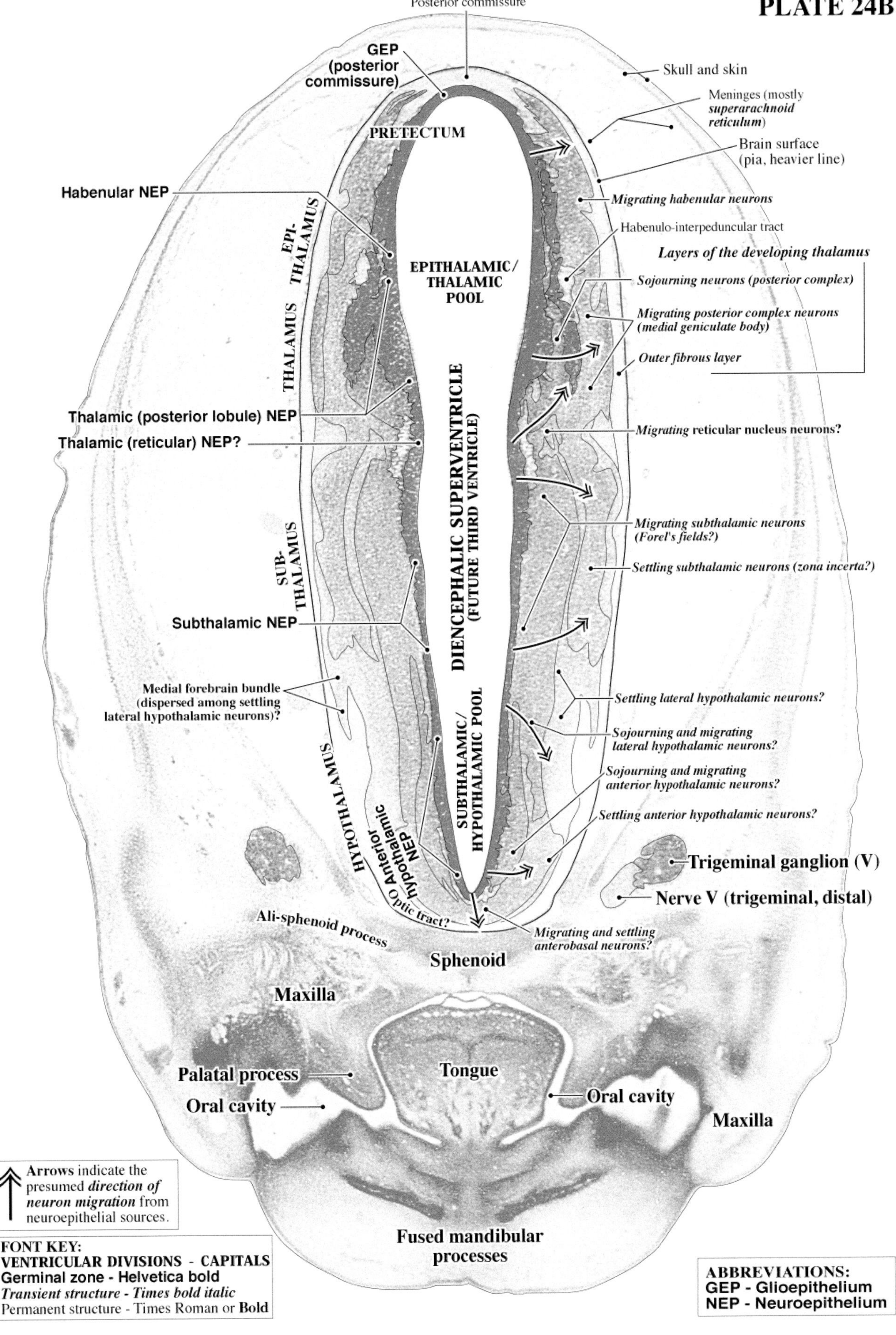

Posterior commissure
GEP (posterior commissure)
PRETECTUM
Skull and skin
Meninges (mostly superarachnoid reticulum)
Brain surface (pia, heavier line)
Habenular NEP
EPI-THALAMUS
Migrating habenular neurons
Habenulo-interpeduncular tract
Layers of the developing thalamus
EPITHALAMIC/ THALAMIC POOL
Sojourning neurons (posterior complex)
Migrating posterior complex neurons (medial geniculate body)
THALAMUS
Outer fibrous layer
Thalamic (posterior lobule) NEP
Thalamic (reticular) NEP?
DIENCEPHALIC SUPERVENTRICLE (FUTURE THIRD VENTRICLE)
Migrating reticular nucleus neurons?
Migrating subthalamic neurons (Forel's fields?)
SUB-THALAMUS
Settling subthalamic neurons (zona incerta?)
Subthalamic NEP
Medial forebrain bundle (dispersed among settling lateral hypothalamic neurons)?
Settling lateral hypothalamic neurons?
Sojourning and migrating lateral hypothalamic neurons?
HYPOTHALAMUS
SUBTHALAMIC/ HYPOTHALAMIC POOL
Sojourning and migrating anterior hypothalamic neurons?
Anterior hypothalamic NEP
Settling anterior hypothalamic neurons?
Trigeminal ganglion (V)
Nerve V (trigeminal, distal)
Optic tract?
Ali-sphenoid process
Migrating and settling anterobasal neurons?
Sphenoid
Maxilla
Tongue
Palatal process
Oral cavity
Oral cavity
Maxilla
Fused mandibular processes

Arrows indicate the presumed *direction of neuron migration* from neuroepithelial sources.

FONT KEY:
VENTRICULAR DIVISIONS - CAPITALS
Germinal zone - Helvetica bold
Transient structure - Times bold italic
Permanent structure - Times Roman or **Bold**

ABBREVIATIONS:
GEP - Glioepithelium
NEP - Neuroepithelium

PLATE 25A

CR 23 mm, GW 8.4, C966
Frontal/horizontal
Level 10: Section 98

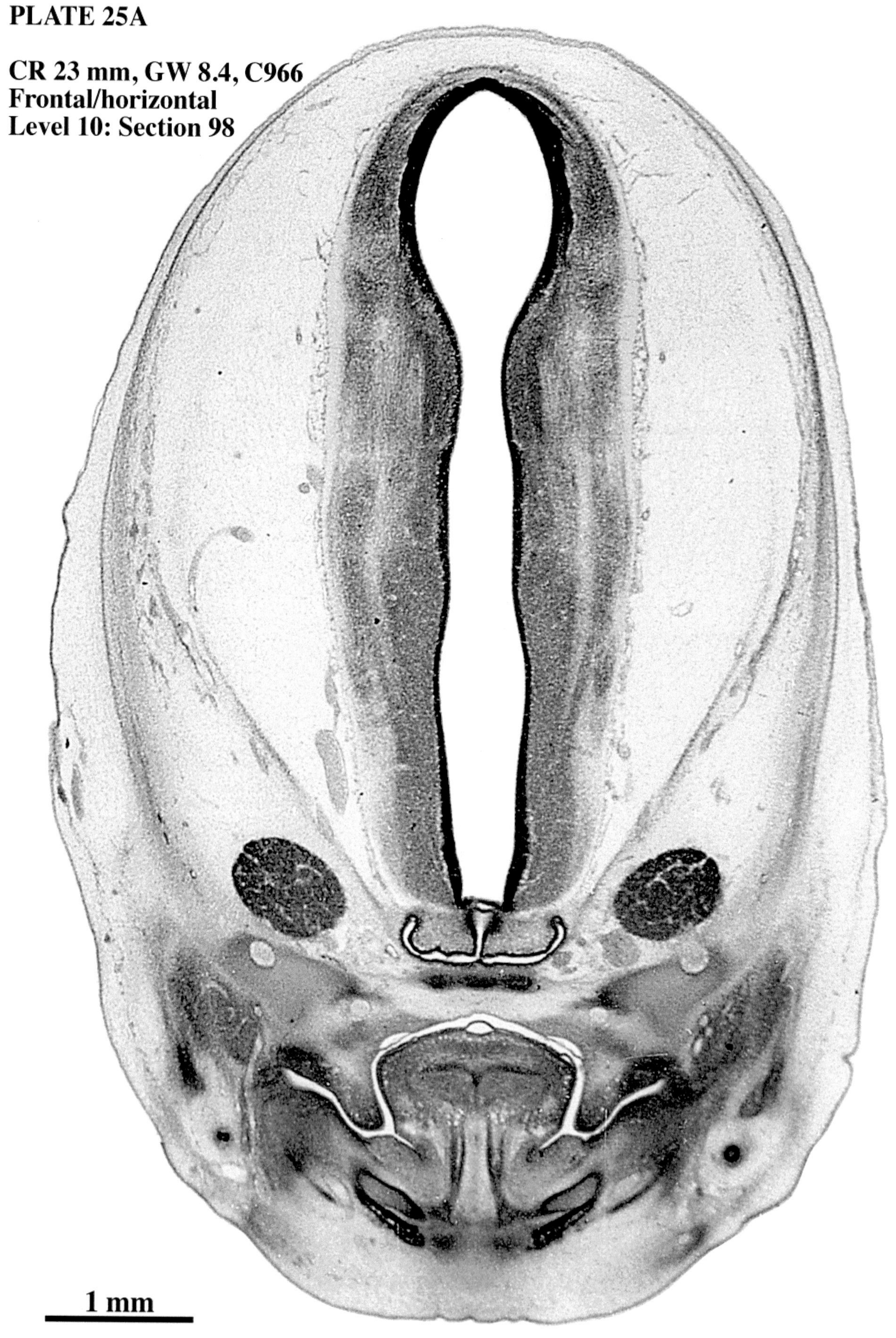
1 mm

PLATE 25B

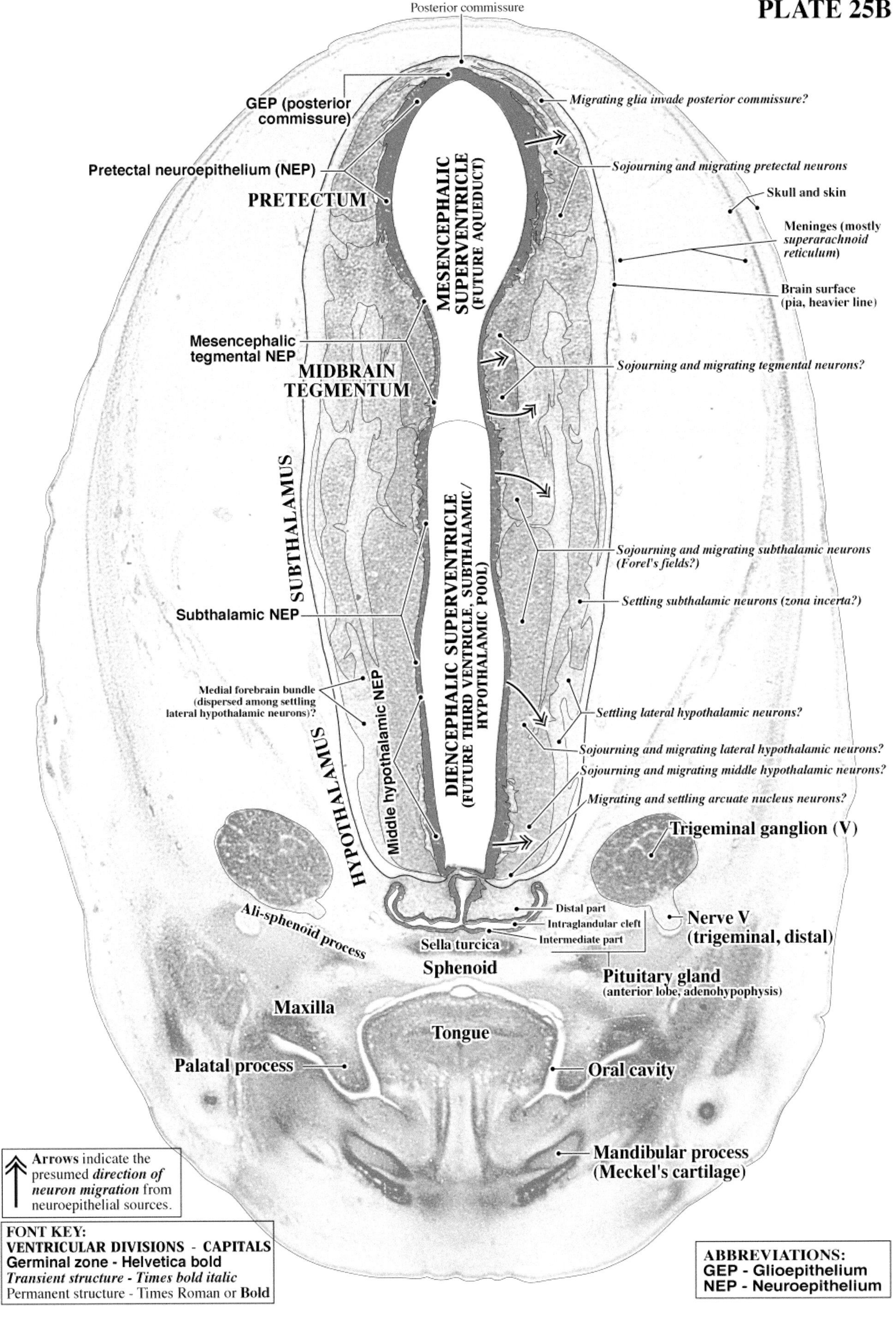
Posterior commissure
GEP (posterior commissure)
Migrating glia invade posterior commissure?
Pretectal neuroepithelium (NEP)
PRETECTUM
MESENCEPHALIC SUPERVENTRICLE (FUTURE AQUEDUCT)
Sojourning and migrating pretectal neurons
Skull and skin
Meninges (mostly superarachnoid reticulum)
Brain surface (pia, heavier line)
Mesencephalic tegmental NEP
MIDBRAIN TEGMENTUM
Sojourning and migrating tegmental neurons?
SUBTHALAMUS
DIENCEPHALIC SUPERVENTRICLE (FUTURE THIRD VENTRICLE, SUBTHALAMIC/ HYPOTHALAMIC POOL)
Sojourning and migrating subthalamic neurons (Forel's fields?)
Settling subthalamic neurons (zona incerta?)
Subthalamic NEP
Medial forebrain bundle (dispersed among settling lateral hypothalamic neurons)?
Middle hypothalamic NEP
HYPOTHALAMUS
Settling lateral hypothalamic neurons?
Sojourning and migrating lateral hypothalamic neurons?
Sojourning and migrating middle hypothalamic neurons?
Migrating and settling arcuate nucleus neurons?
Trigeminal ganglion (V)
Distal part
Intraglandular cleft
Intermediate part
Nerve V (trigeminal, distal)
Ali-sphenoid process
Sella turcica
Sphenoid
Pituitary gland (anterior lobe, adenohypophysis)
Maxilla
Tongue
Palatal process
Oral cavity
Mandibular process (Meckel's cartilage)
Arrows indicate the presumed direction of neuron migration from neuroepithelial sources.
FONT KEY:
VENTRICULAR DIVISIONS - CAPITALS
Germinal zone - Helvetica bold
Transient structure - Times bold italic
Permanent structure - Times Roman or Bold
ABBREVIATIONS:
GEP - Glioepithelium
NEP - Neuroepithelium

PLATE 26A

CR 23 mm, GW 8.4, C966
Frontal/horizontal
Level 11: Section 105

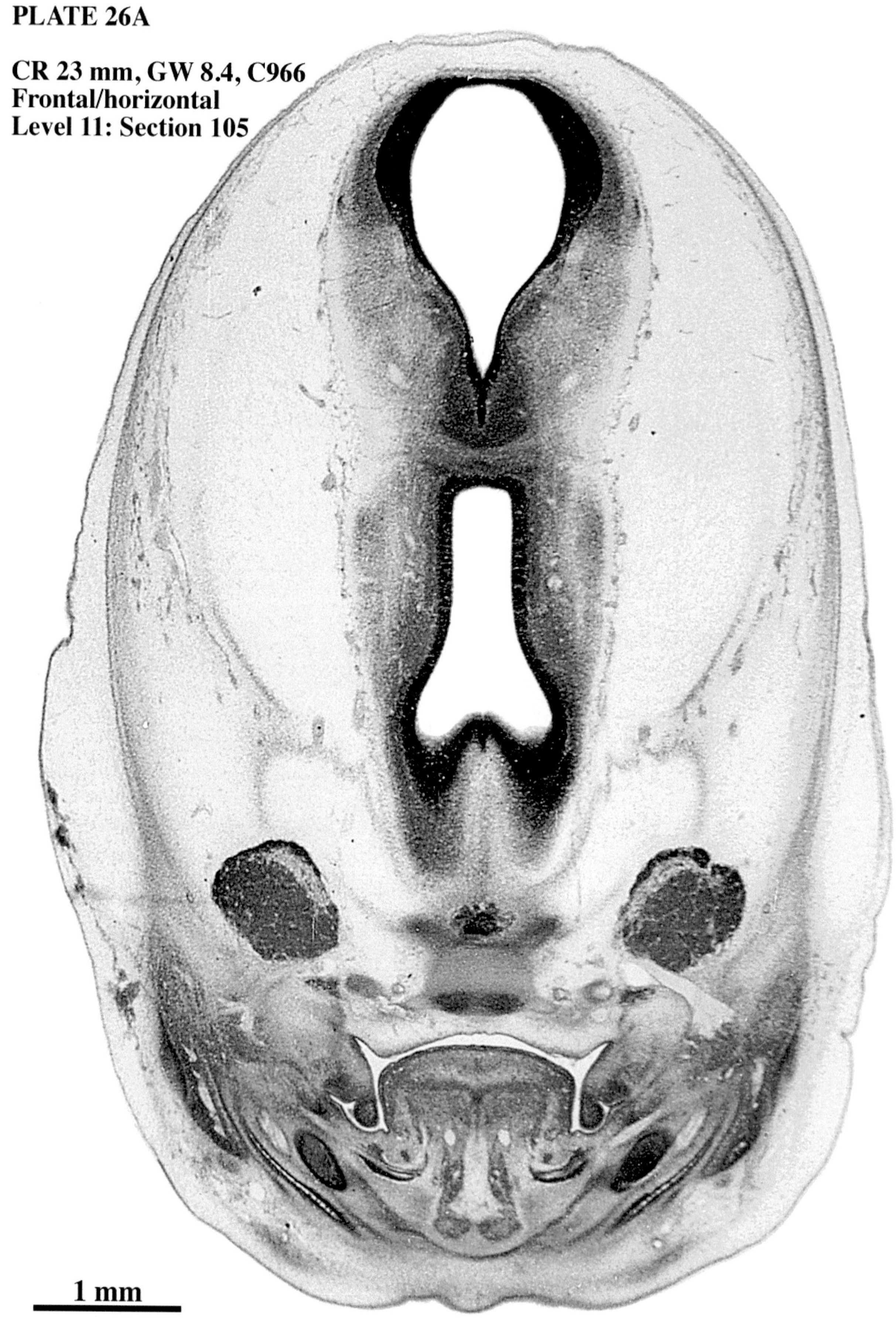
1 mm

PLATE 26B

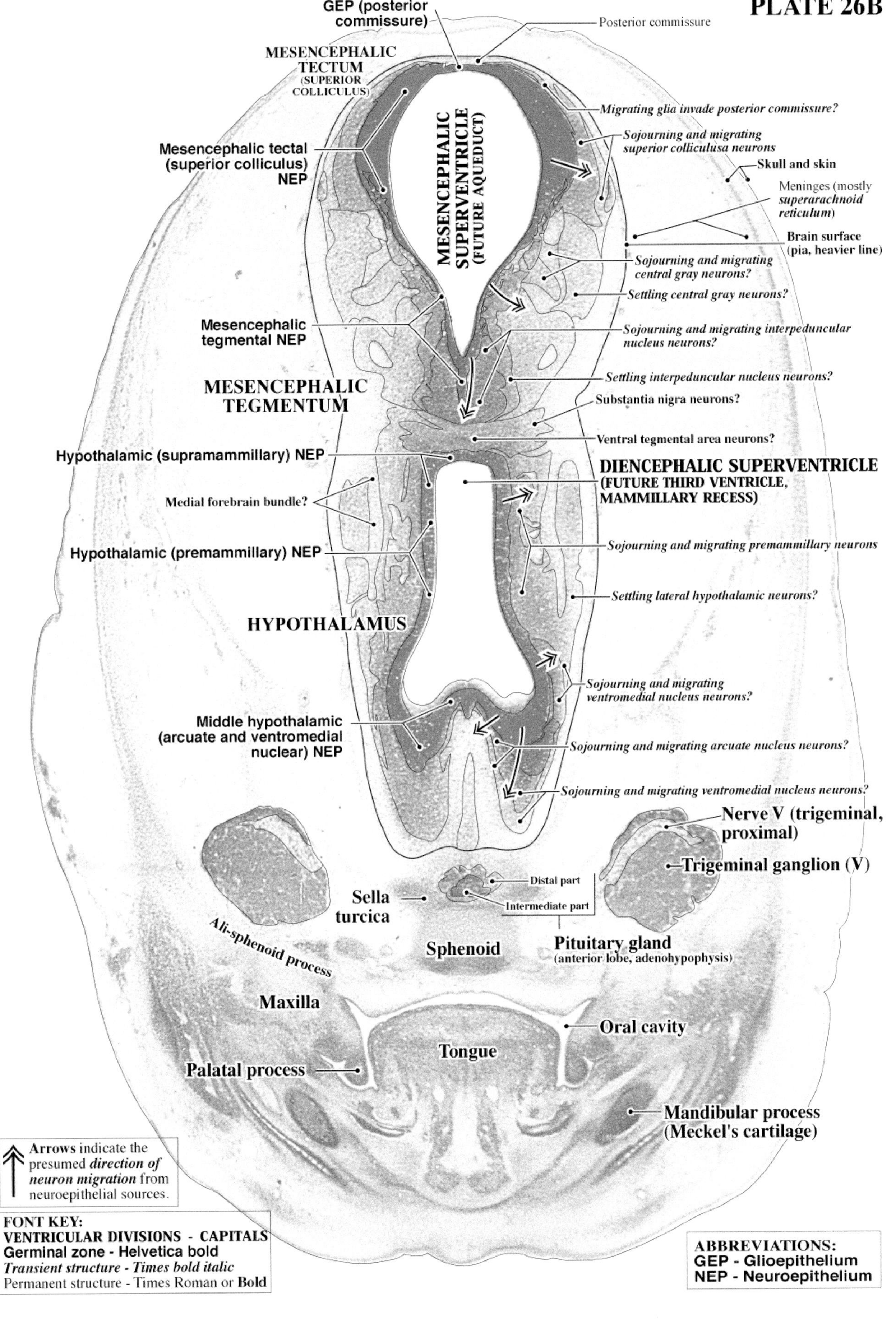

GEP (posterior commissure)
Posterior commissure
MESENCEPHALIC TECTUM (SUPERIOR COLLICULUS)
MESENCEPHALIC SUPERVENTRICLE (FUTURE AQUEDUCT)
Migrating glia invade posterior commissure?
Mesencephalic tectal (superior colliculus) NEP
Sojourning and migrating superior colliculusa neurons
Skull and skin
Meninges (mostly superarachnoid reticulum)
Brain surface (pia, heavier line)
Sojourning and migrating central gray neurons?
Settling central gray neurons?
Mesencephalic tegmental NEP
Sojourning and migrating interpeduncular nucleus neurons?
Settling interpeduncular nucleus neurons?
MESENCEPHALIC TEGMENTUM
Substantia nigra neurons?
Ventral tegmental area neurons?
Hypothalamic (supramammillary) NEP
DIENCEPHALIC SUPERVENTRICLE (FUTURE THIRD VENTRICLE, MAMMILLARY RECESS)
Medial forebrain bundle?
Hypothalamic (premammillary) NEP
Sojourning and migrating premammillary neurons
Settling lateral hypothalamic neurons?
HYPOTHALAMUS
Sojourning and migrating ventromedial nucleus neurons?
Middle hypothalamic (arcuate and ventromedial nuclear) NEP
Sojourning and migrating arcuate nucleus neurons?
Sojourning and migrating ventromedial nucleus neurons?
Nerve V (trigeminal, proximal)
Trigeminal ganglion (V)
Distal part
Intermediate part
Sella turcica
Ali-sphenoid process
Sphenoid
Pituitary gland (anterior lobe, adenohypophysis)
Maxilla
Oral cavity
Tongue
Palatal process
Mandibular process (Meckel's cartilage)
Arrows indicate the presumed direction of neuron migration from neuroepithelial sources.
FONT KEY:
VENTRICULAR DIVISIONS - CAPITALS
Germinal zone - Helvetica bold
Transient structure - Times bold italic
Permanent structure - Times Roman or Bold
ABBREVIATIONS:
GEP - Glioepithelium
NEP - Neuroepithelium

PLATE 27A

CR 23 mm, GW 8.4, C966
Frontal/horizontal
Level 12: Section 112

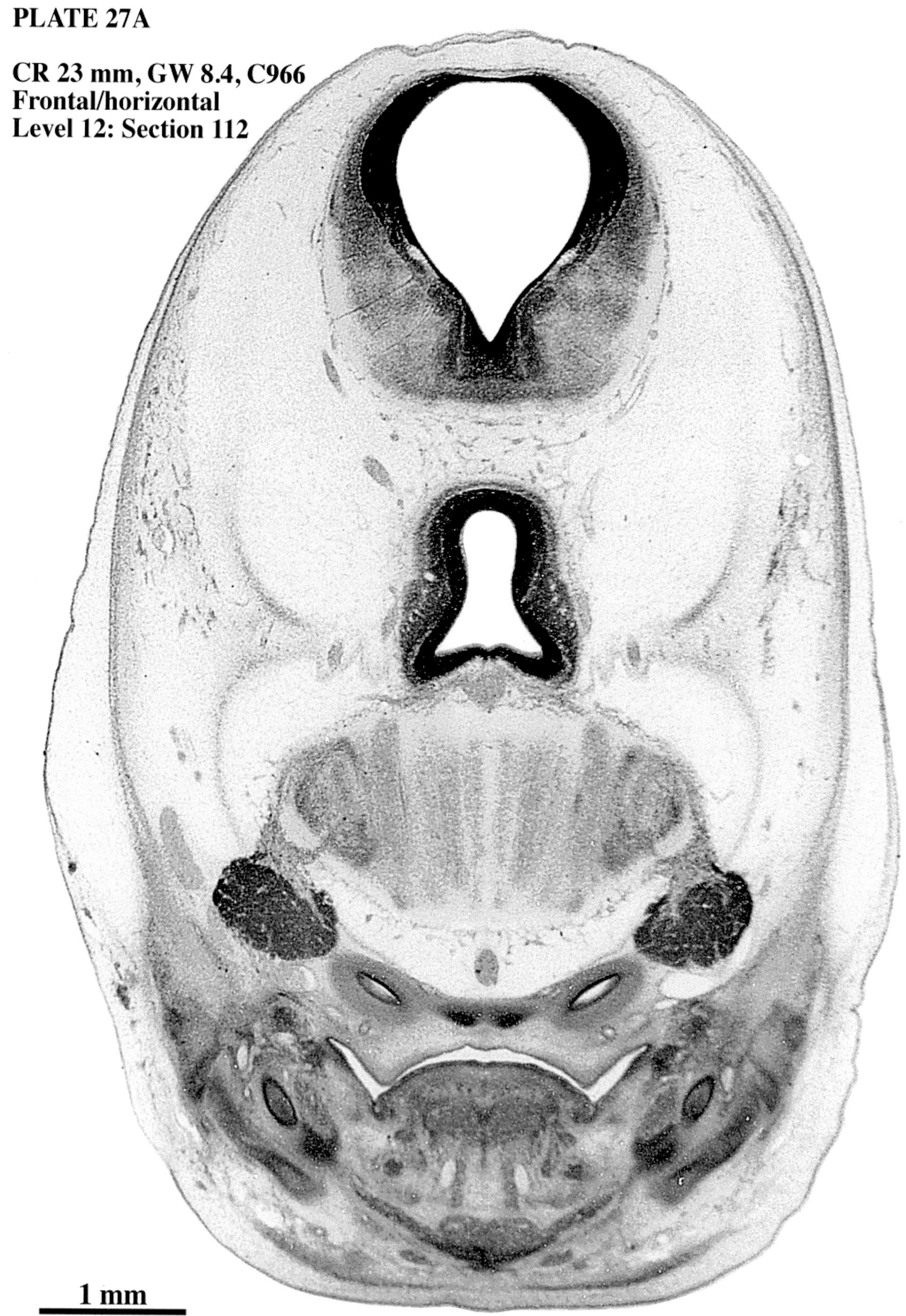
1 mm

PLATE 27B

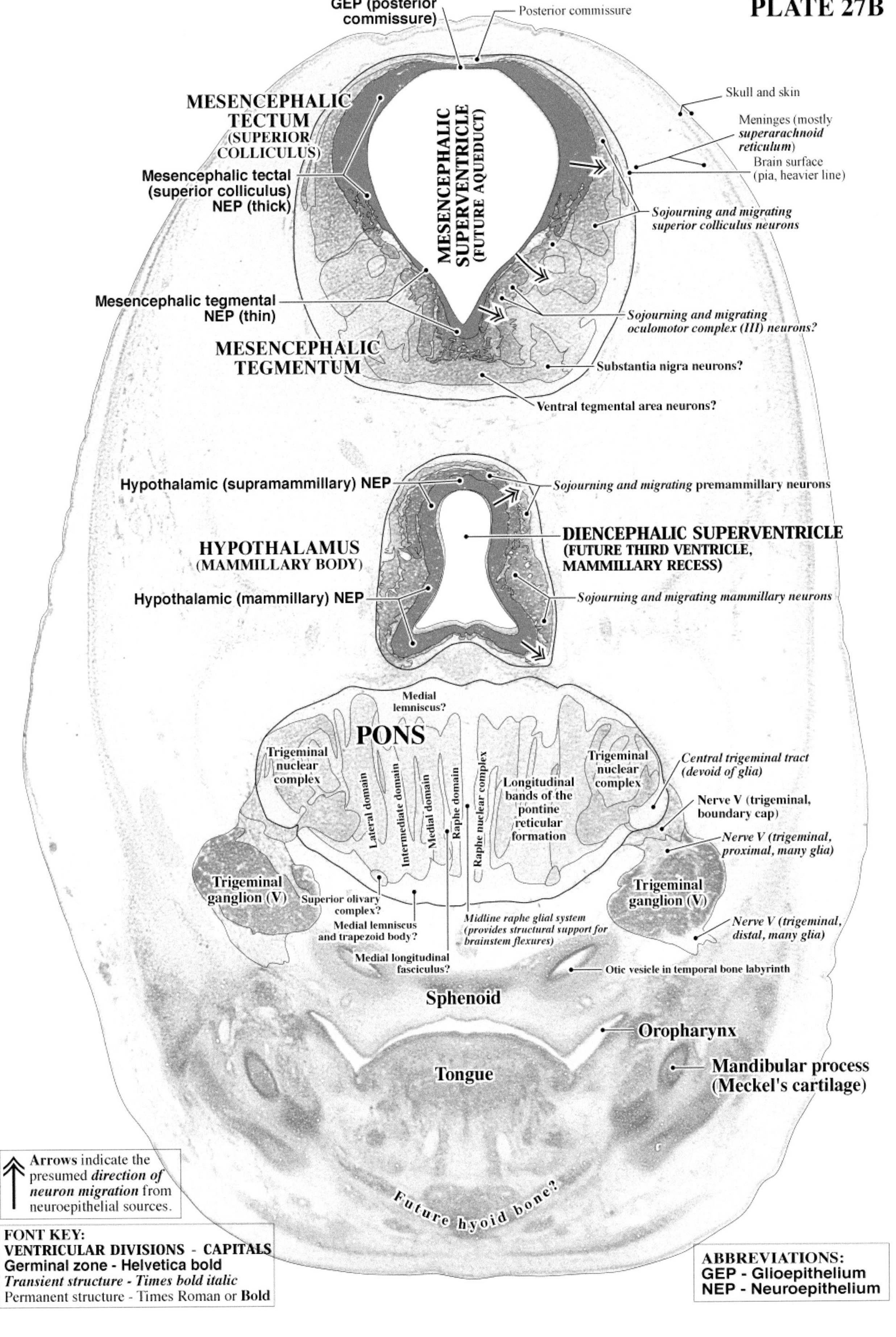
GEP (posterior commissure)
Posterior commissure
Skull and skin
Meninges (mostly superarachnoid reticulum)
Brain surface (pia, heavier line)
MESENCEPHALIC TECTUM (SUPERIOR COLLICULUS)
Mesencephalic tectal (superior colliculus) NEP (thick)
MESENCEPHALIC SUPERVENTRICLE (FUTURE AQUEDUCT)
Sojourning and migrating superior colliculus neurons
Mesencephalic tegmental NEP (thin)
Sojourning and migrating oculomotor complex (III) neurons?
MESENCEPHALIC TEGMENTUM
Substantia nigra neurons?
Ventral tegmental area neurons?
Hypothalamic (supramammillary) NEP
Sojourning and migrating premammillary neurons
HYPOTHALAMUS (MAMMILLARY BODY)
DIENCEPHALIC SUPERVENTRICLE (FUTURE THIRD VENTRICLE, MAMMILLARY RECESS)
Hypothalamic (mammillary) NEP
Sojourning and migrating mammillary neurons
Medial lemniscus?
PONS
Trigeminal nuclear complex
Lateral domain
Intermediate domain
Medial domain
Raphe domain
Raphe nuclear complex
Longitudinal bands of the pontine reticular formation
Trigeminal nuclear complex
Central trigeminal tract (devoid of glia)
Nerve V (trigeminal, boundary cap)
Nerve V (trigeminal, proximal, many glia)
Trigeminal ganglion (V)
Superior olivary complex?
Medial lemniscus and trapezoid body?
Medial longitudinal fasciculus?
Midline raphe glial system (provides structural support for brainstem flexures)
Trigeminal ganglion (V)
Nerve V (trigeminal, distal, many glia)
Otic vesicle in temporal bone labyrinth
Sphenoid
Oropharynx
Tongue
Mandibular process (Meckel's cartilage)
Future hyoid bone?
Arrows indicate the presumed direction of neuron migration from neuroepithelial sources.
FONT KEY:
VENTRICULAR DIVISIONS - CAPITALS
Germinal zone - Helvetica bold
Transient structure - Times bold italic
Permanent structure - Times Roman or Bold
ABBREVIATIONS:
GEP - Glioepithelium
NEP - Neuroepithelium

PLATE 28A

CR 23 mm, GW 8.4, C966
Frontal/horizontal
Level 13: Section 119

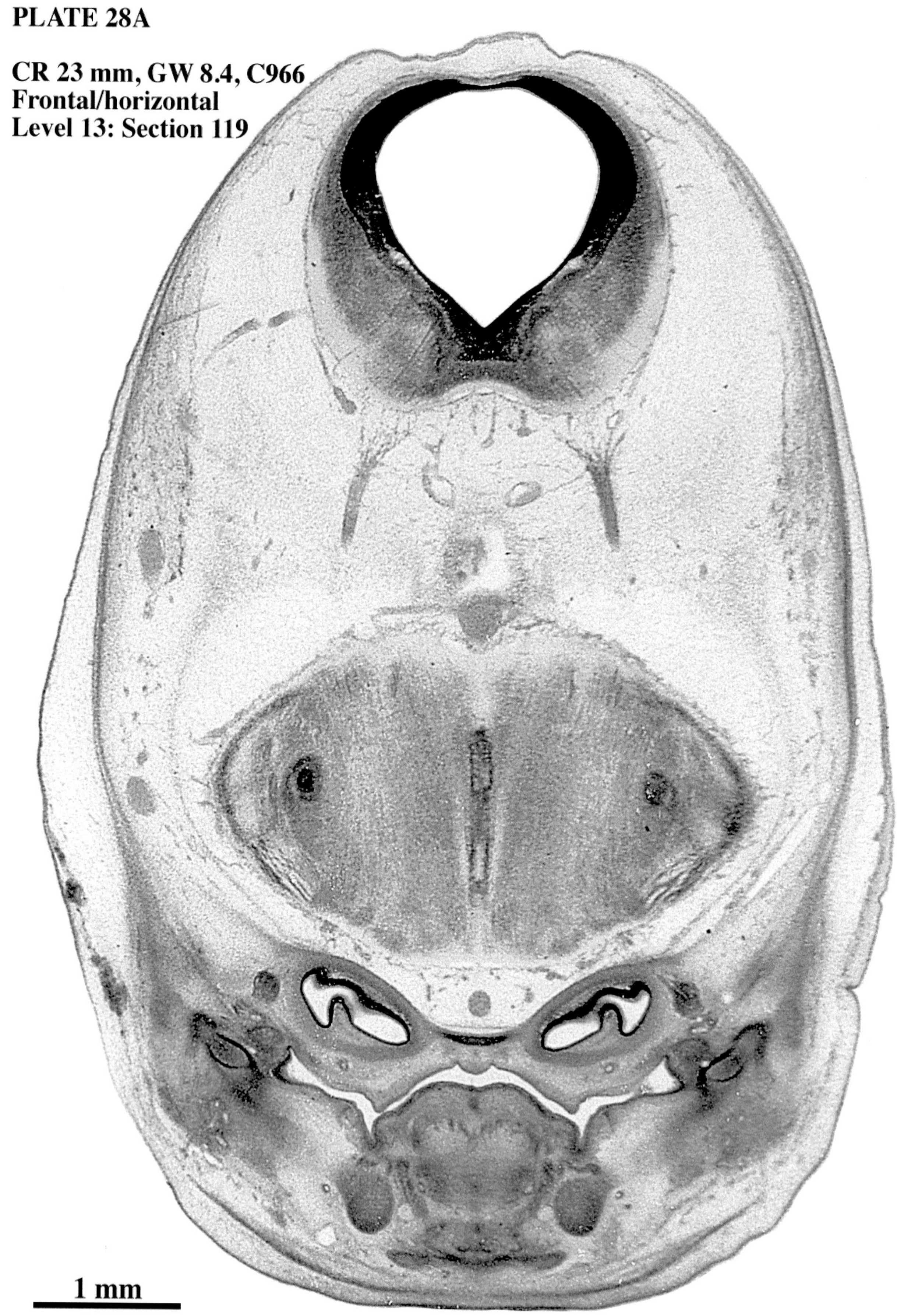
1 mm

PLATE 28B

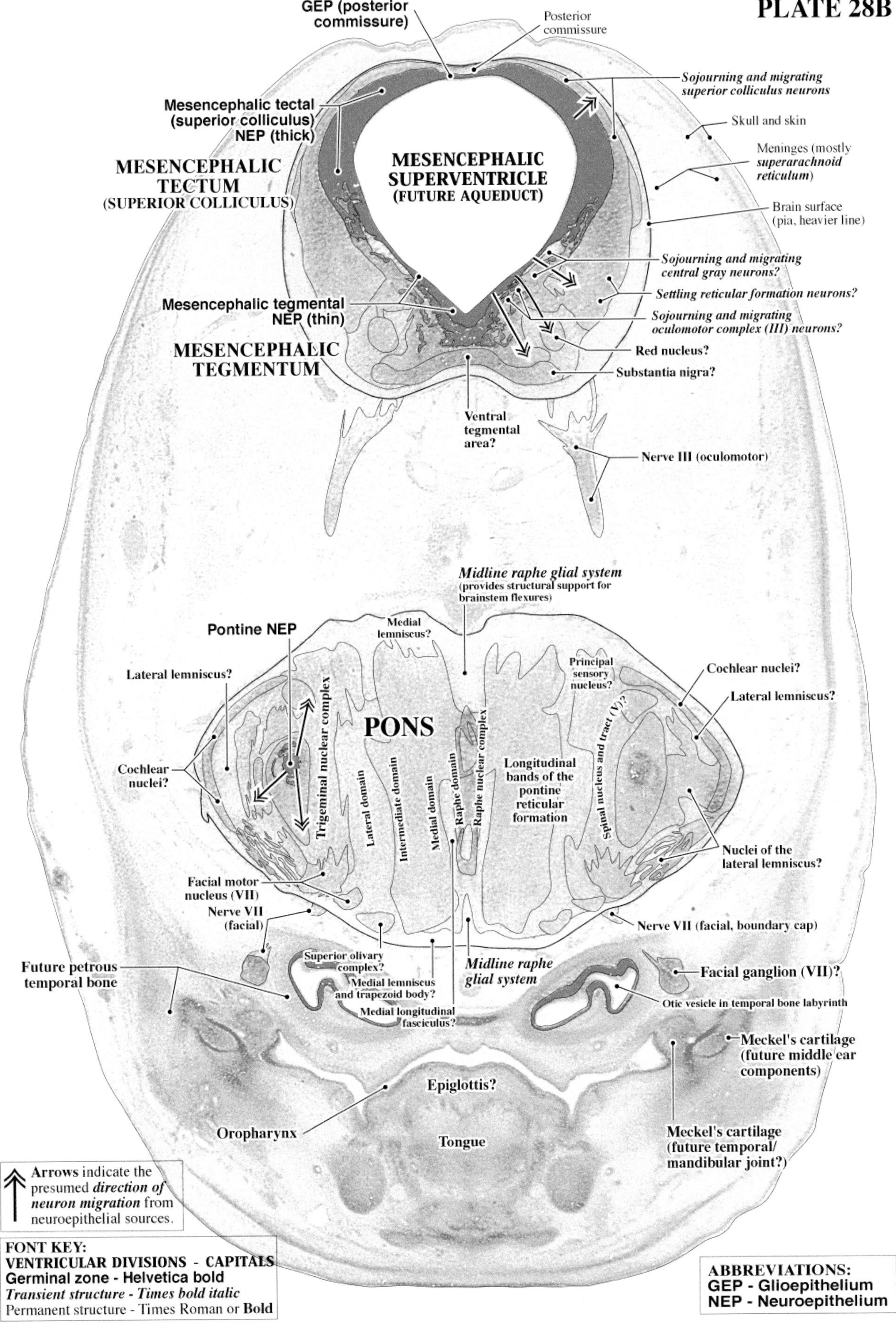

GEP (posterior commissure)
Posterior commissure
Sojourning and migrating superior colliculus neurons
Skull and skin
Meninges (mostly superarachnoid reticulum)
Brain surface (pia, heavier line)
Mesencephalic tectal (superior colliculus) NEP (thick)
MESENCEPHALIC TECTUM (SUPERIOR COLLICULUS)
MESENCEPHALIC SUPERVENTRICLE (FUTURE AQUEDUCT)
Sojourning and migrating central gray neurons?
Settling reticular formation neurons?
Sojourning and migrating oculomotor complex (III) neurons?
Mesencephalic tegmental NEP (thin)
MESENCEPHALIC TEGMENTUM
Red nucleus?
Substantia nigra?
Ventral tegmental area?
Nerve III (oculomotor)
Midline raphe glial system
(provides structural support for brainstem flexures)
Pontine NEP
Medial lemniscus?
Lateral lemniscus?
Principal sensory nucleus?
Cochlear nuclei?
Lateral lemniscus?
PONS
Trigeminal nuclear complex
Lateral domain
Intermediate domain
Medial domain
Raphe domain
Raphe nuclear complex
Longitudinal bands of the pontine reticular formation
Spinal nucleus and tract (V)?
Cochlear nuclei?
Nuclei of the lateral lemniscus?
Facial motor nucleus (VII)
Nerve VII (facial)
Nerve VII (facial, boundary cap)
Future petrous temporal bone
Superior olivary complex?
Medial lemniscus and trapezoid body?
Medial longitudinal fasciculus?
Midline raphe glial system
Facial ganglion (VII)?
Otic vesicle in temporal bone labyrinth
Meckel's cartilage (future middle ear components)
Epiglottis?
Oropharynx
Tongue
Meckel's cartilage (future temporal/ mandibular joint?)
Arrows indicate the presumed direction of neuron migration from neuroepithelial sources.
FONT KEY:
VENTRICULAR DIVISIONS - CAPITALS
Germinal zone - Helvetica bold
Transient structure - Times bold italic
Permanent structure - Times Roman or Bold
ABBREVIATIONS:
GEP - Glioepithelium
NEP - Neuroepithelium

PLATE 29A

CR 23 mm, GW 8.4, C966
Frontal/horizontal
Level 14: Section 125

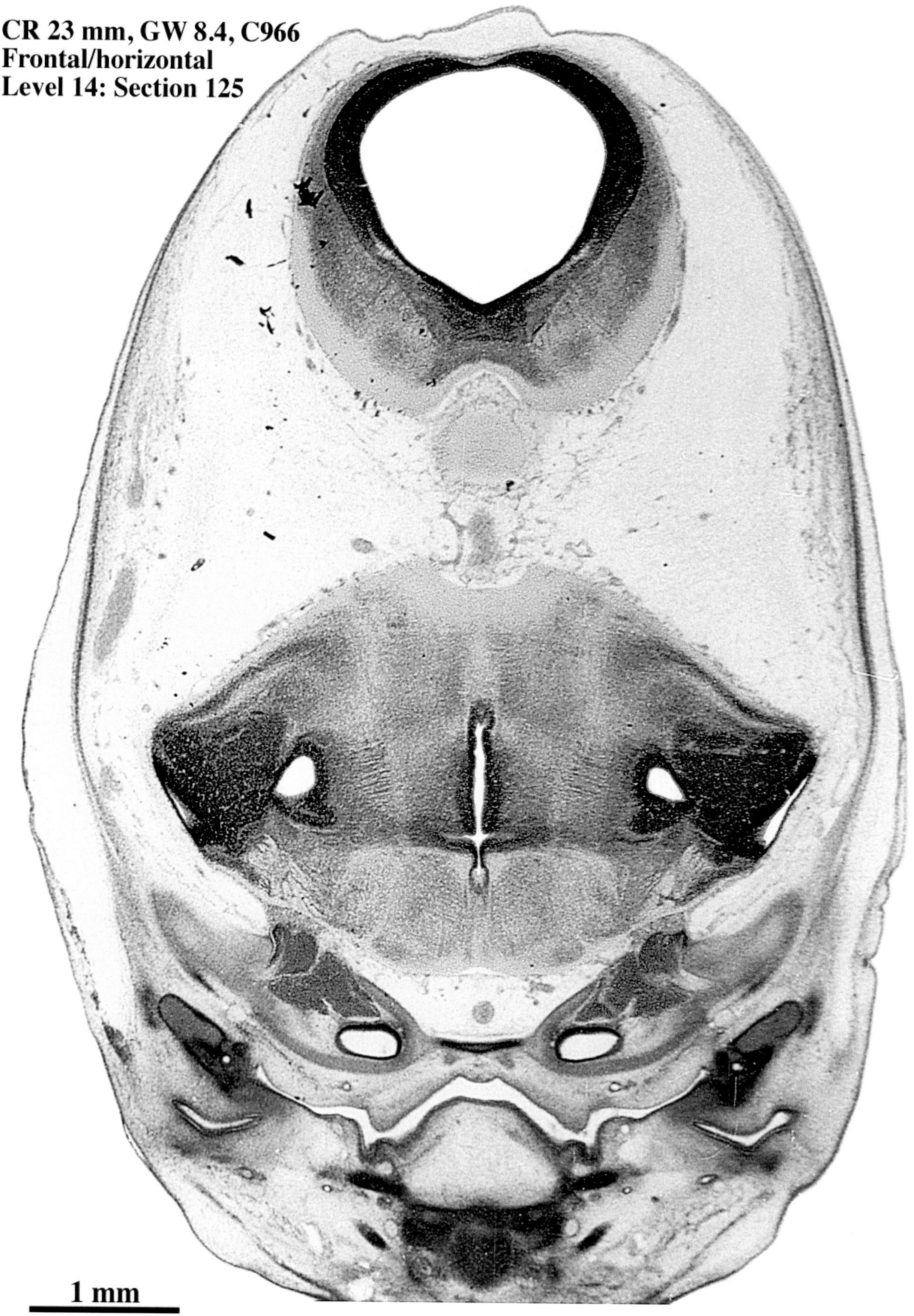

1 mm

PLATE 29B

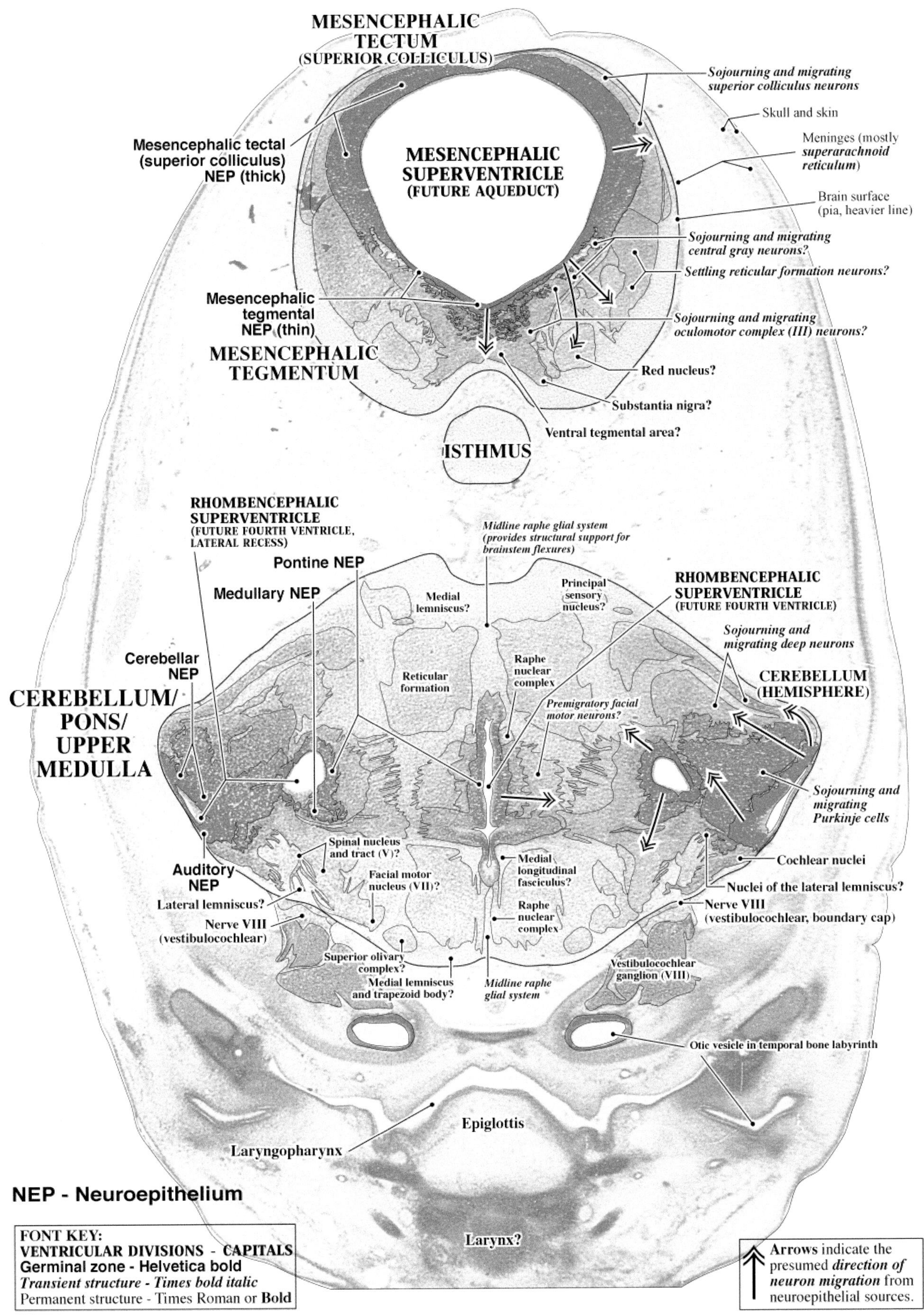
MESENCEPHALIC TECTUM (SUPERIOR COLLICULUS)
Sojourning and migrating superior colliculus neurons
Skull and skin
Meninges (mostly superarachnoid reticulum)
Mesencephalic tectal (superior colliculus) NEP (thick)
MESENCEPHALIC SUPERVENTRICLE (FUTURE AQUEDUCT)
Brain surface (pia, heavier line)
Sojourning and migrating central gray neurons?
Settling reticular formation neurons?
Mesencephalic tegmental NEP (thin)
Sojourning and migrating oculomotor complex (III) neurons?
MESENCEPHALIC TEGMENTUM
Red nucleus?
Substantia nigra?
Ventral tegmental area?
ISTHMUS
RHOMBENCEPHALIC SUPERVENTRICLE (FUTURE FOURTH VENTRICLE, LATERAL RECESS)
Midline raphe glial system (provides structural support for brainstem flexures)
Pontine NEP
Medullary NEP
Medial lemniscus?
Principal sensory nucleus?
RHOMBENCEPHALIC SUPERVENTRICLE (FUTURE FOURTH VENTRICLE)
Sojourning and migrating deep neurons
Cerebellar NEP
Reticular formation
Raphe nuclear complex
CEREBELLUM (HEMISPHERE)
CEREBELLUM/ PONS/ UPPER MEDULLA
Premigratory facial motor neurons?
Sojourning and migrating Purkinje cells
Spinal nucleus and tract (V)?
Medial longitudinal fasciculus?
Cochlear nuclei
Auditory NEP
Facial motor nucleus (VII)?
Nuclei of the lateral lemniscus?
Lateral lemniscus?
Raphe nuclear complex
Nerve VIII (vestibulocochlear, boundary cap)
Nerve VIII (vestibulocochlear)
Superior olivary complex?
Vestibulocochlear ganglion (VIII)
Medial lemniscus and trapezoid body?
Midline raphe glial system
Otic vesicle in temporal bone labyrinth
Epiglottis
Laryngopharynx
NEP - Neuroepithelium
Larynx?
FONT KEY:
VENTRICULAR DIVISIONS - CAPITALS
Germinal zone - Helvetica bold
Transient structure - Times bold italic
Permanent structure - Times Roman or Bold
Arrows indicate the presumed direction of neuron migration from neuroepithelial sources.

PLATE 30A

CR 23 mm, GW 8.4, C966
Frontal/horizontal
Level 15: Section 128

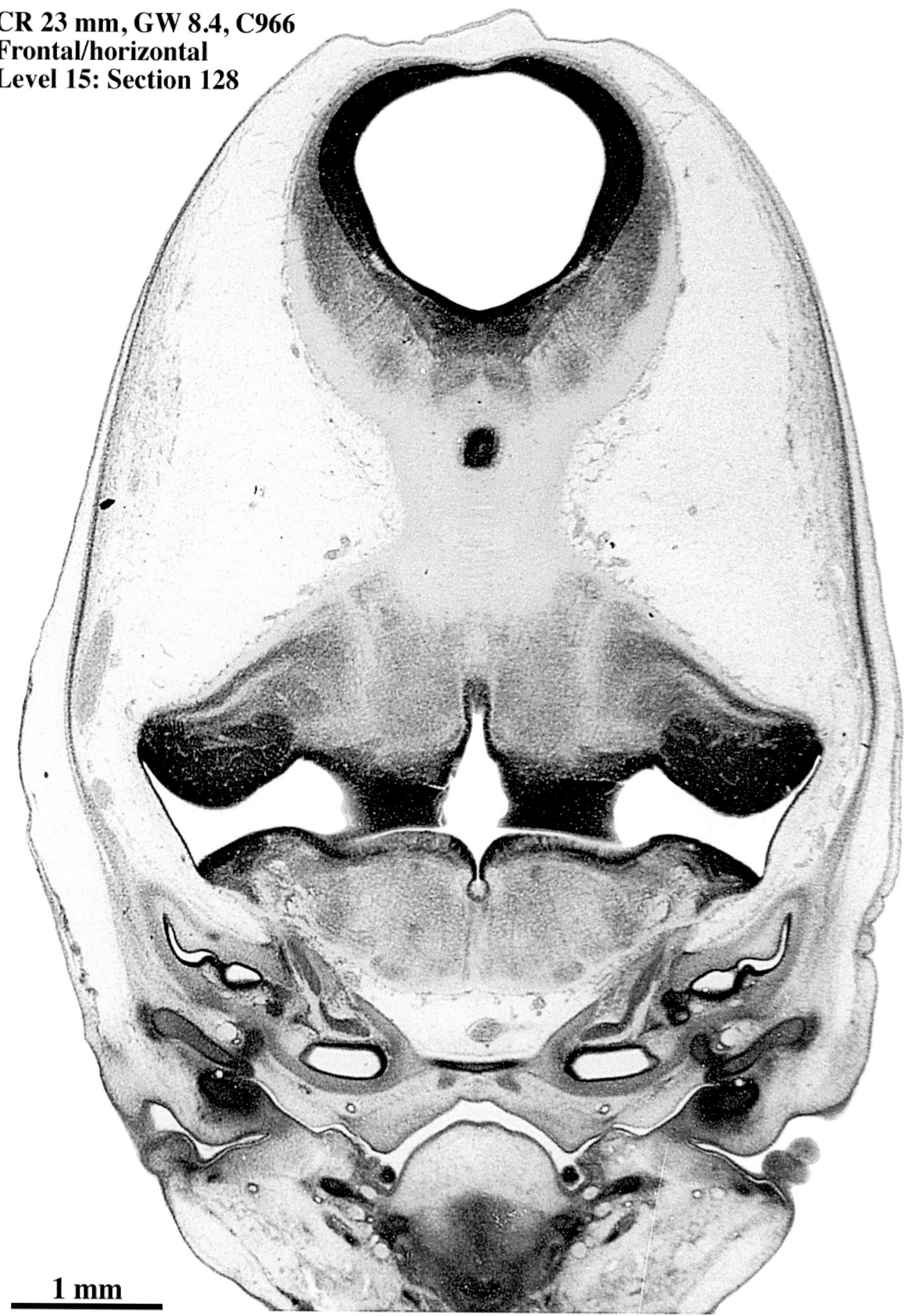
1 mm

PLATE 30B

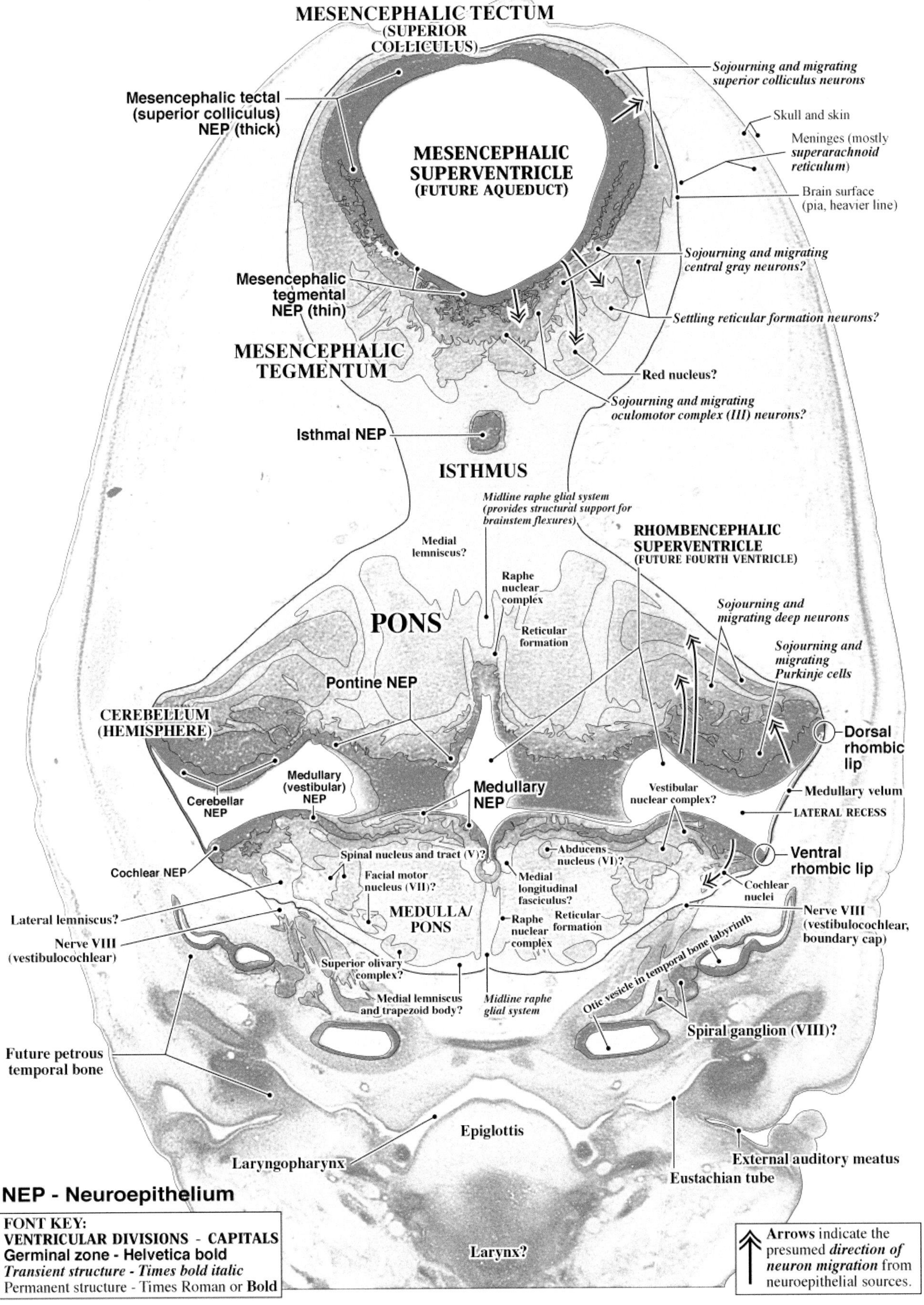
MESENCEPHALIC TECTUM
(SUPERIOR COLLICULUS)
Mesencephalic tectal (superior colliculus) NEP (thick)
Sojourning and migrating superior colliculus neurons
Skull and skin
Meninges (mostly superarachnoid reticulum)
Brain surface (pia, heavier line)
MESENCEPHALIC SUPERVENTRICLE (FUTURE AQUEDUCT)
Sojourning and migrating central gray neurons?
Mesencephalic tegmental NEP (thin)
Settling reticular formation neurons?
MESENCEPHALIC TEGMENTUM
Red nucleus?
Sojourning and migrating oculomotor complex (III) neurons?
Isthmal NEP
ISTHMUS
Midline raphe glial system (provides structural support for brainstem flexures)
Medial lemniscus?
RHOMBENCEPHALIC SUPERVENTRICLE (FUTURE FOURTH VENTRICLE)
Raphe nuclear complex
PONS
Reticular formation
Sojourning and migrating deep neurons
Sojourning and migrating Purkinje cells
Pontine NEP
CEREBELLUM (HEMISPHERE)
Dorsal rhombic lip
Medullary (vestibular) NEP
Medullary NEP
Vestibular nuclear complex?
Medullary velum
Cerebellar NEP
LATERAL RECESS
Abducens nucleus (VI)?
Spinal nucleus and tract (V)?
Ventral rhombic lip
Cochlear NEP
Facial motor nucleus (VII)?
Medial longitudinal fasciculus?
Cochlear nuclei
MEDULLA/ PONS
Reticular formation
Raphe nuclear complex
Lateral lemniscus?
Nerve VIII (vestibulocochlear, boundary cap)
Nerve VIII (vestibulocochlear)
Otic vesicle in temporal bone labyrinth
Superior olivary complex?
Medial lemniscus and trapezoid body?
Midline raphe glial system
Spiral ganglion (VIII)?
Future petrous temporal bone
Epiglottis
Laryngopharynx
External auditory meatus
Eustachian tube
Larynx?

NEP - Neuroepithelium

FONT KEY:
VENTRICULAR DIVISIONS - CAPITALS
Germinal zone - Helvetica bold
Transient structure - Times bold italic
Permanent structure - Times Roman or **Bold**

Arrows indicate the presumed ***direction of neuron migration*** from neuroepithelial sources.

PLATE 31A

CR 23 mm, GW 8.4, C966
Frontal/horizontal
Level 16: Section 139

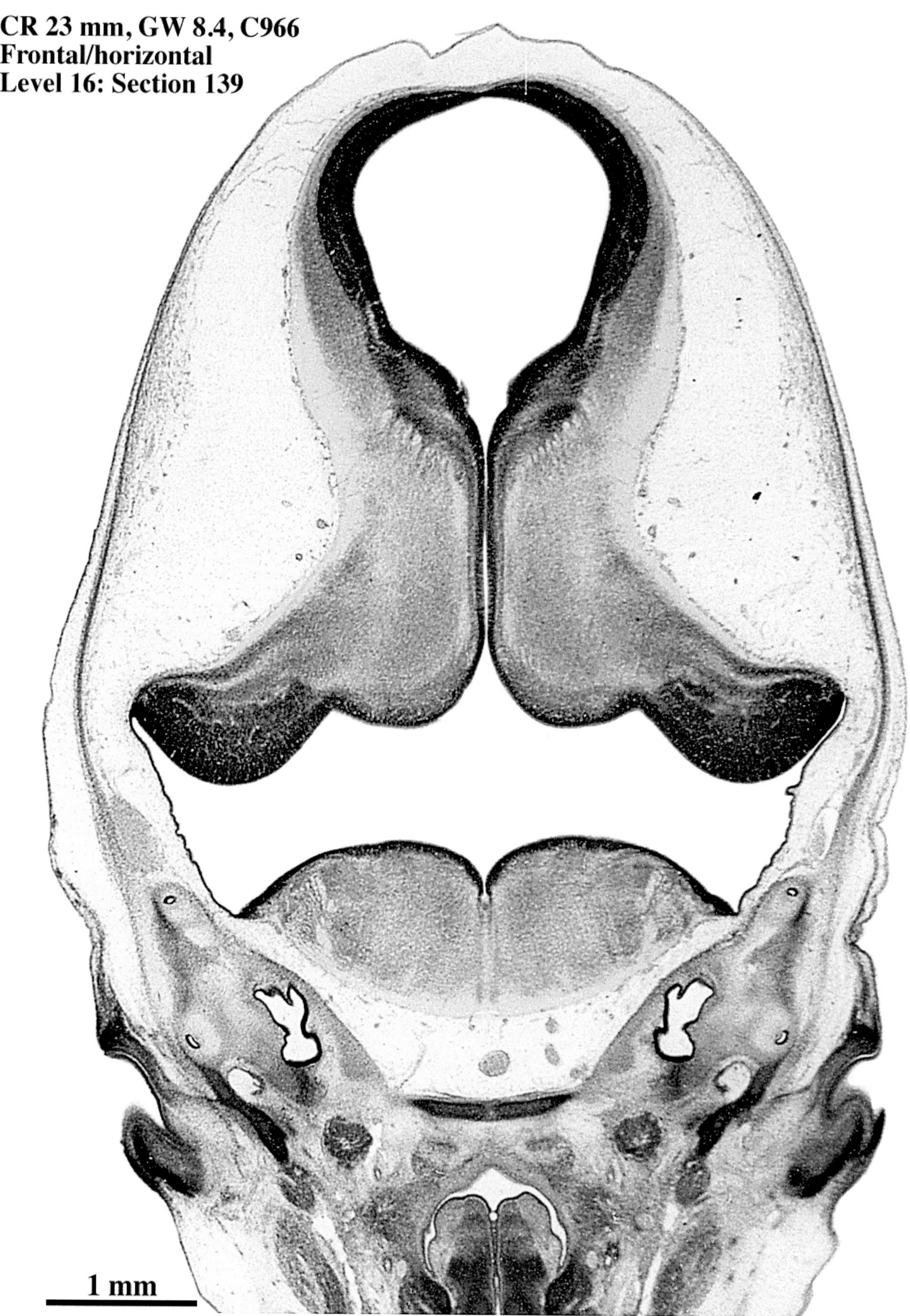
1 mm

PLATE 31B

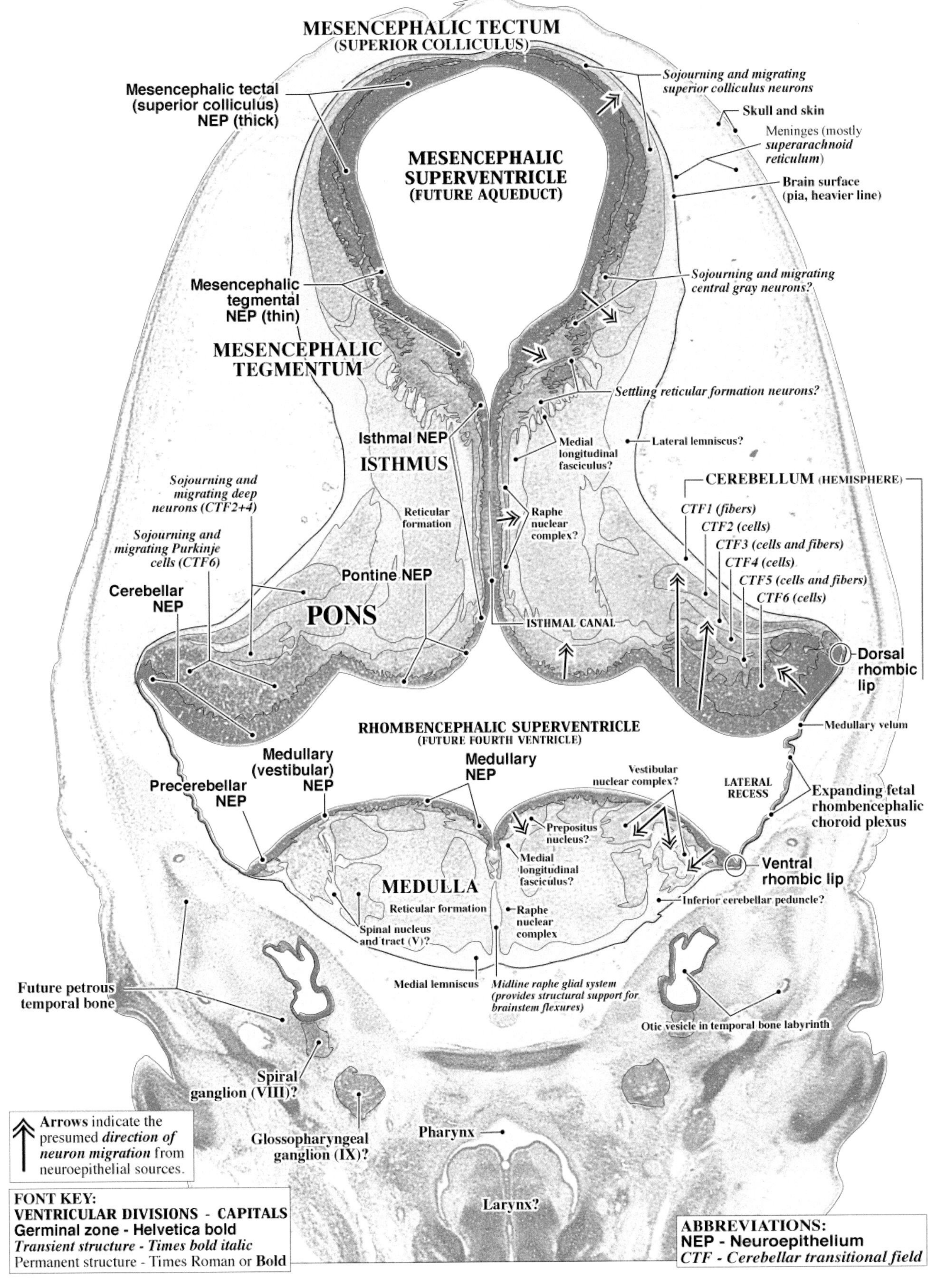
MESENCEPHALIC TECTUM
(SUPERIOR COLLICULUS)
Mesencephalic tectal (superior colliculus) NEP (thick)
Sojourning and migrating superior colliculus neurons
Skull and skin
Meninges (mostly superarachnoid reticulum)
Brain surface (pia, heavier line)
MESENCEPHALIC SUPERVENTRICLE (FUTURE AQUEDUCT)
Mesencephalic tegmental NEP (thin)
Sojourning and migrating central gray neurons?
MESENCEPHALIC TEGMENTUM
Settling reticular formation neurons?
Isthmal NEP
ISTHMUS
Medial longitudinal fasciculus?
Lateral lemniscus?
Sojourning and migrating deep neurons (CTF2+4)
Sojourning and migrating Purkinje cells (CTF6)
Reticular formation
Raphe nuclear complex?
CEREBELLUM (HEMISPHERE)
CTF1 (fibers)
CTF2 (cells)
CTF3 (cells and fibers)
CTF4 (cells)
CTF5 (cells and fibers)
CTF6 (cells)
Cerebellar NEP
Pontine NEP
PONS
ISTHMAL CANAL
Dorsal rhombic lip
RHOMBENCEPHALIC SUPERVENTRICLE
(FUTURE FOURTH VENTRICLE)
Medullary velum
Medullary (vestibular) NEP
Medullary NEP
Precerebellar NEP
Vestibular nuclear complex?
LATERAL RECESS
Expanding fetal rhombencephalic choroid plexus
Prepositus nucleus?
Medial longitudinal fasciculus?
Ventral rhombic lip
MEDULLA
Reticular formation
Raphe nuclear complex
Inferior cerebellar peduncle?
Spinal nucleus and tract (V)?
Medial lemniscus
Midline raphe glial system (provides structural support for brainstem flexures)
Future petrous temporal bone
Otic vesicle in temporal bone labyrinth
Spiral ganglion (VIII)?
Glossopharyngeal ganglion (IX)?
Pharynx
Larynx?
Arrows indicate the presumed direction of neuron migration from neuroepithelial sources.
FONT KEY:
VENTRICULAR DIVISIONS - CAPITALS
Germinal zone - Helvetica bold
Transient structure - Times bold italic
Permanent structure - Times Roman or Bold
ABBREVIATIONS:
NEP - Neuroepithelium
CTF - Cerebellar transitional field

PLATE 32A

CR 23 mm, GW 8.4, C966
Frontal/horizontal
Level 17: Section 147

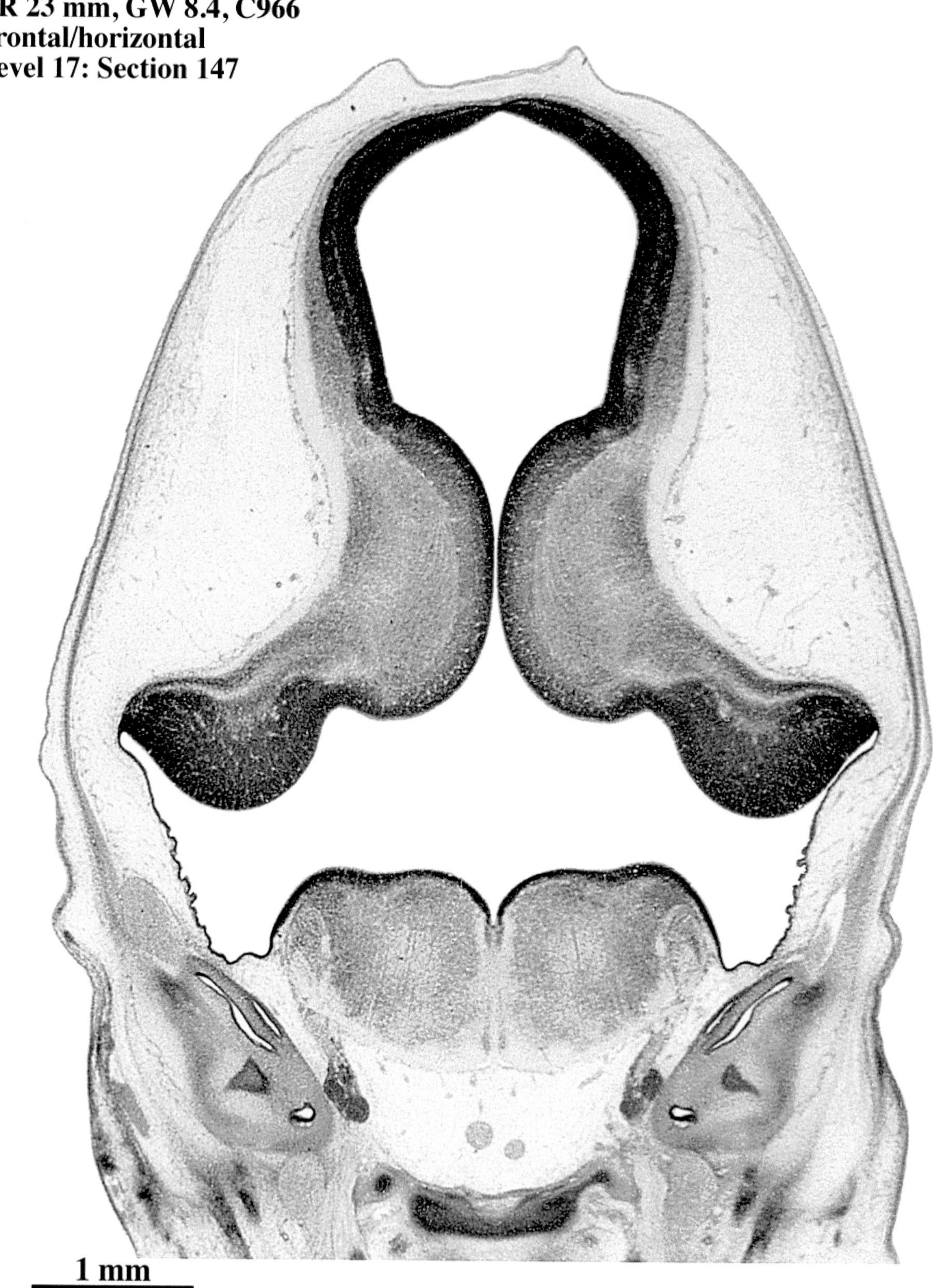

1 mm

PLATE 32B

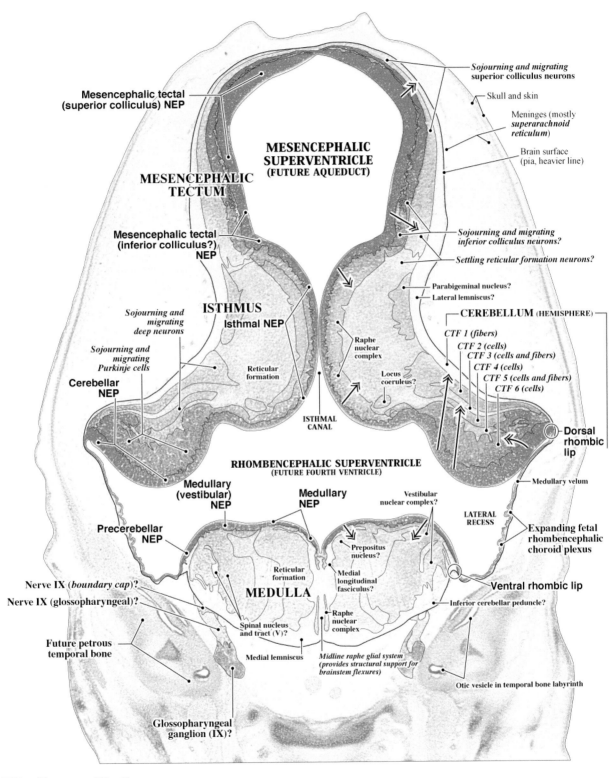

Sojourning and migrating superior colliculus neurons

Skull and skin

Meninges (mostly *superarachnoid reticulum*)

Brain surface (pia, heavier line)

Mesencephalic tectal (superior colliculus) NEP

MESENCEPHALIC SUPERVENTRICLE (FUTURE AQUEDUCT)

MESENCEPHALIC TECTUM

Mesencephalic tectal (inferior colliculus?) NEP

Sojourning and migrating inferior colliculus neurons?

Settling reticular formation neurons?

Parabigeminal nucleus?

Lateral lemniscus?

CEREBELLUM (HEMISPHERE)

CTF 1 (fibers)

CTF 2 (cells)

CTF 3 (cells and fibers)

CTF 4 (cells)

CTF 5 (cells and fibers)

CTF 6 (cells)

ISTHMUS

Isthmal NEP

Sojourning and migrating deep neurons

Sojourning and migrating Purkinje cells

Cerebellar NEP

Raphe nuclear complex

Reticular formation

Locus coeruleus?

ISTHMAL CANAL

Dorsal rhombic lip

RHOMBENCEPHALIC SUPERVENTRICLE (FUTURE FOURTH VENTRICLE)

Medullary velum

Medullary (vestibular) NEP

Medullary NEP

Vestibular nuclear complex?

LATERAL RECESS

Precerebellar NEP

Expanding fetal rhombencephalic choroid plexus

Prepositus nucleus?

Reticular formation

Medial longitudinal fasciculus?

Ventral rhombic lip

Nerve IX (*boundary cap*)?

Nerve IX (glossopharyngeal)?

MEDULLA

Inferior cerebellar peduncle?

Raphe nuclear complex

Spinal nucleus and tract (V)?

Future petrous temporal bone

Medial lemniscus

Midline raphe glial system (provides structural support for brainstem flexures)

Otic vesicle in temporal bone labyrinth

Glossopharyngeal ganglion (IX)?

NEP - Neuroepithelium

FONT KEY:
VENTRICULAR DIVISIONS - CAPITALS
Germinal zone - Helvetica bold
Transient structure - Times bold italic
Permanent structure - Times Roman or **Bold**

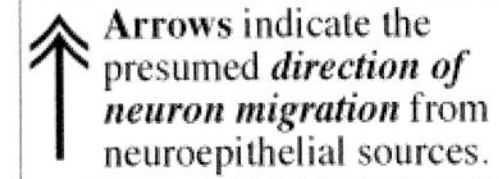

Arrows indicate the presumed ***direction of neuron migration*** from neuroepithelial sources.

PLATE 33A

CR 23 mm, GW 8.4, C966
Frontal/horizontal
Level 18: Section 157

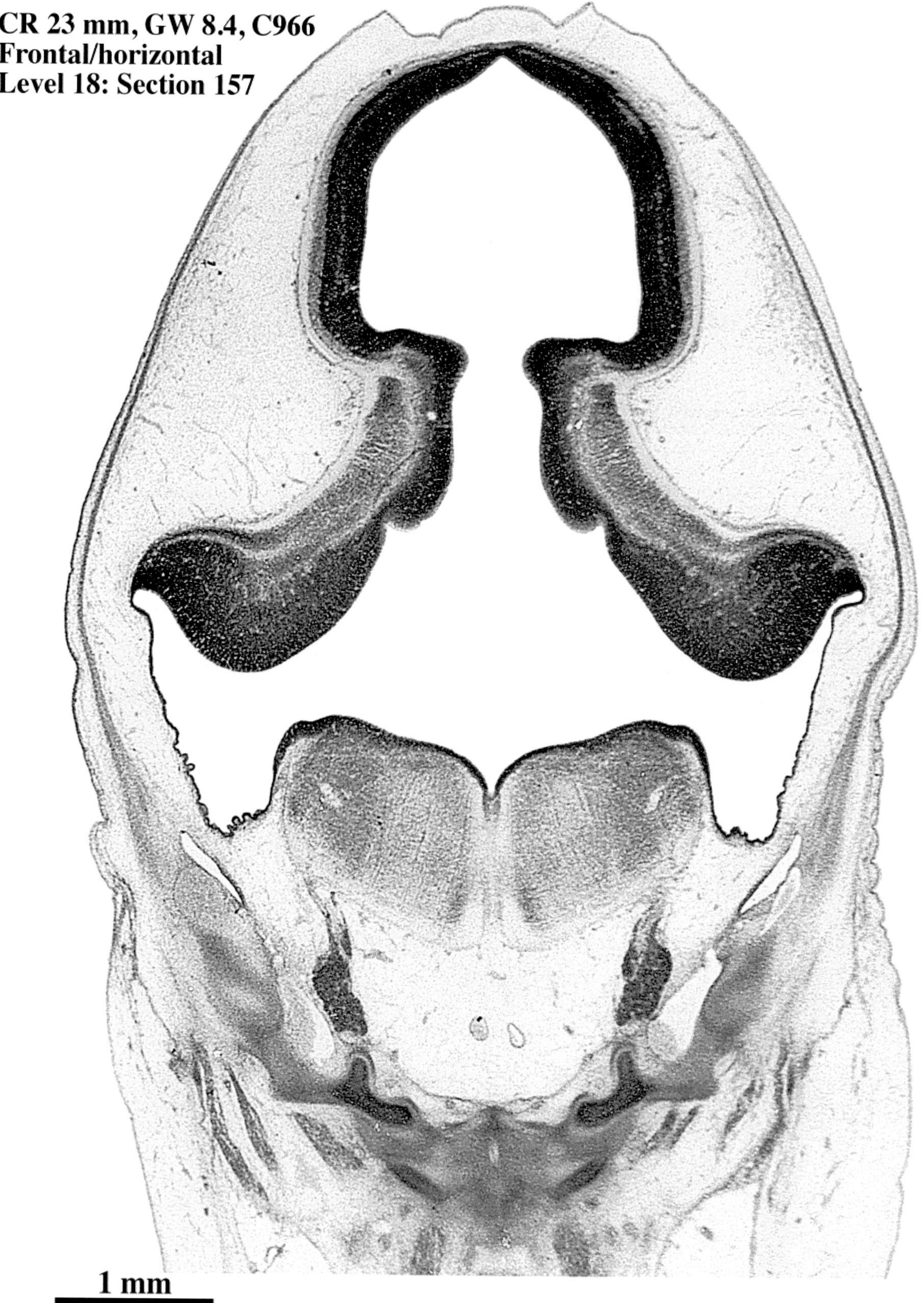
1 mm

PLATE 33B

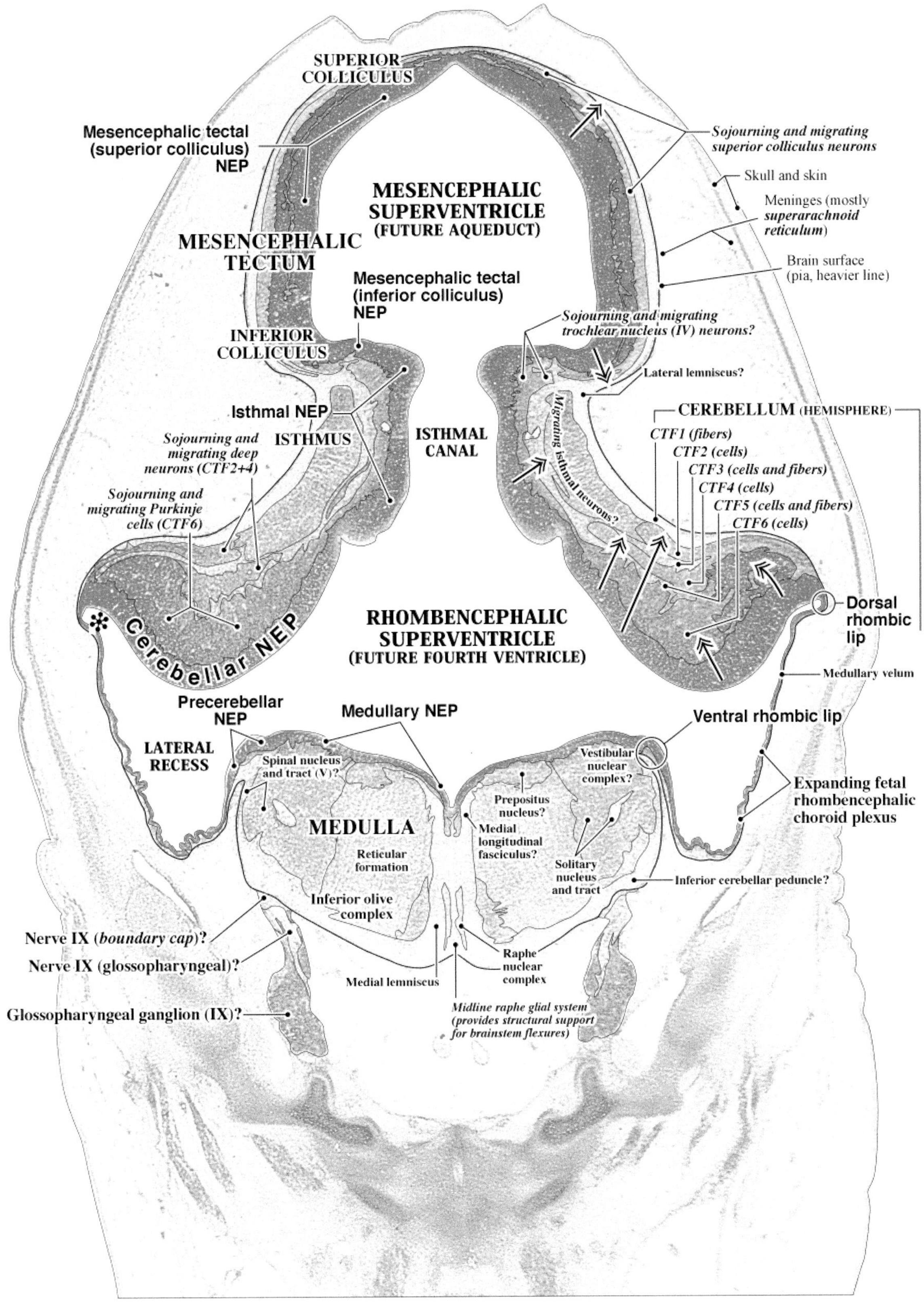

SUPERIOR COLLICULUS
Mesencephalic tectal (superior colliculus) NEP
Sojourning and migrating superior colliculus neurons
Skull and skin
Meninges (mostly *superarachnoid reticulum*)
Brain surface (pia, heavier line)
MESENCEPHALIC SUPERVENTRICLE (FUTURE AQUEDUCT)
MESENCEPHALIC TECTUM
Mesencephalic tectal (inferior colliculus) NEP
INFERIOR COLLICULUS
Sojourning and migrating trochlear nucleus (IV) neurons?
Lateral lemniscus?
Isthmal NEP
ISTHMUS
ISTHMAL CANAL
Migrating isthmal neurons?
Sojourning and migrating deep neurons (CTF2+4)
Sojourning and migrating Purkinje cells (CTF6)
CEREBELLUM (HEMISPHERE)
CTF1 (fibers)
CTF2 (cells)
CTF3 (cells and fibers)
CTF4 (cells)
CTF5 (cells and fibers)
CTF6 (cells)
Cerebellar NEP
RHOMBENCEPHALIC SUPERVENTRICLE (FUTURE FOURTH VENTRICLE)
Dorsal rhombic lip
Medullary velum
Precerebellar NEP
Medullary NEP
Ventral rhombic lip
LATERAL RECESS
Spinal nucleus and tract (V)?
Vestibular nuclear complex?
Prepositus nucleus?
Expanding fetal rhombencephalic choroid plexus
MEDULLA
Medial longitudinal fasciculus?
Reticular formation
Solitary nucleus and tract
Inferior cerebellar peduncle?
Inferior olive complex
Nerve IX (*boundary cap*)?
Nerve IX (glossopharyngeal)?
Raphe nuclear complex
Medial lemniscus
Midline raphe glial system (provides structural support for brainstem flexures)
Glossopharyngeal ganglion (IX)?

FONT KEY:
VENTRICULAR DIVISIONS - CAPITALS
Germinal zone - Helvetica bold
Transient structure - Times bold italic
Permanent structure - Times Roman or **Bold**

ABBREVIATIONS:
NEP - Neuroepithelium
CTF - Cerebellar transitional field

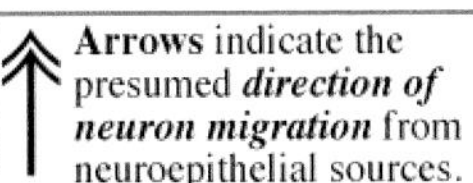

Arrows indicate the presumed ***direction of neuron migration*** from neuroepithelial sources.

PLATE 34A

CR 23 mm, GW 8.4, C966
Frontal/horizontal
Level 19: Section 163

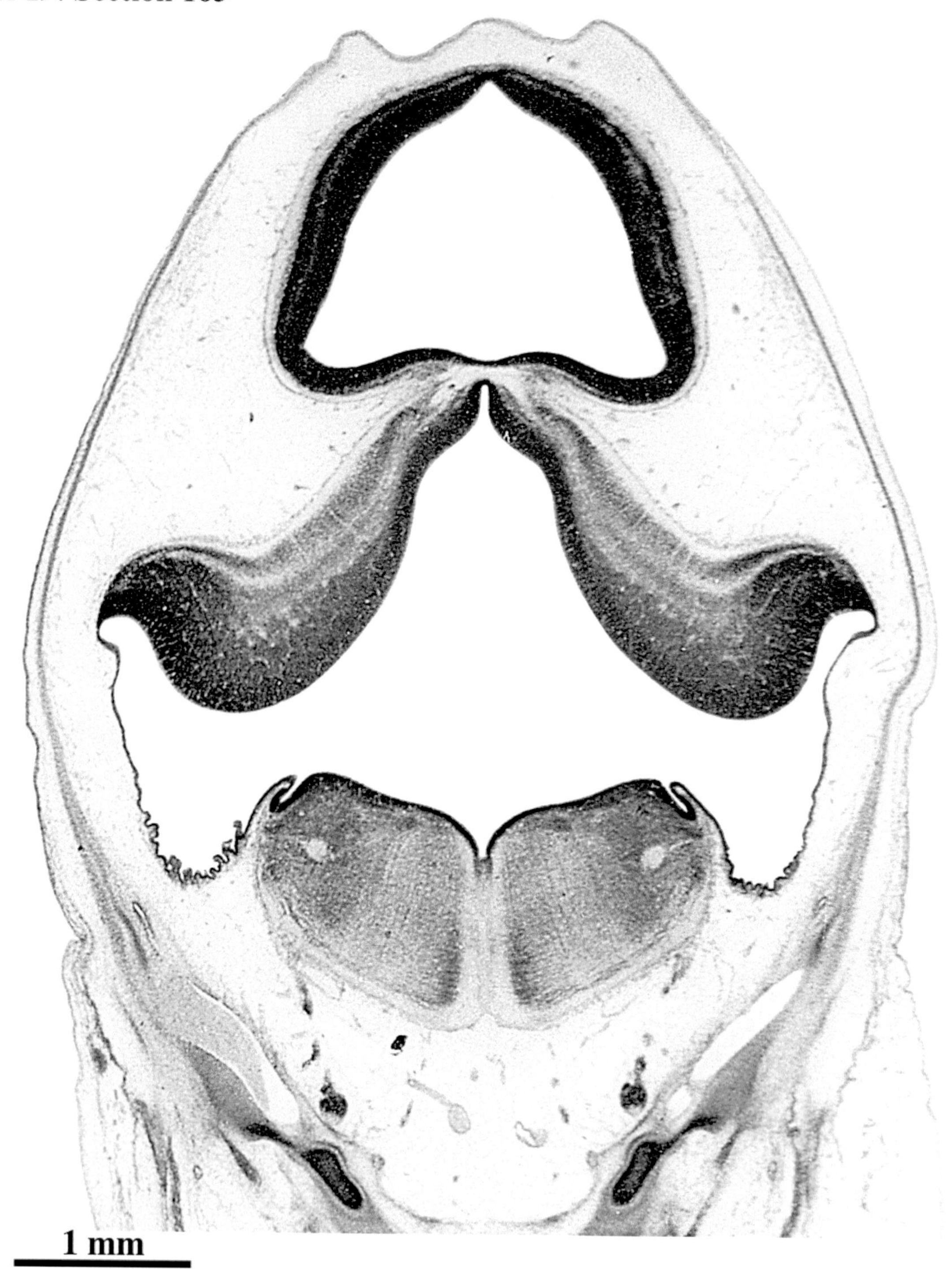
1 mm

PLATE 34B

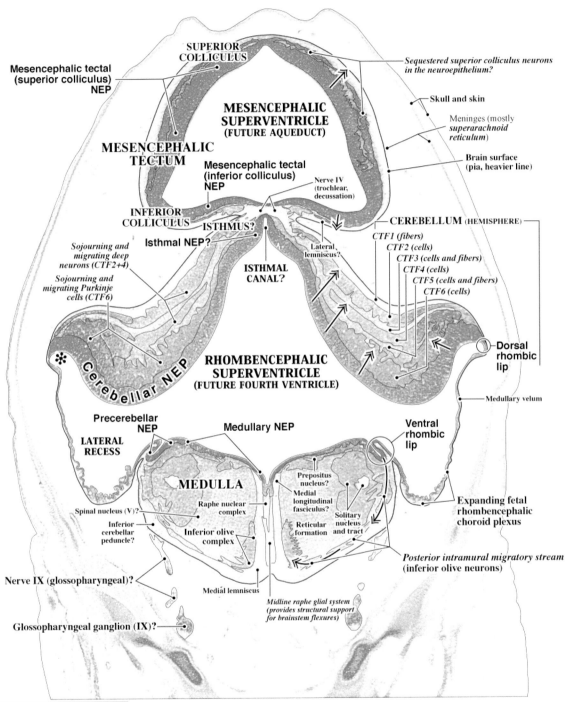

SUPERIOR COLLICULUS

Mesencephalic tectal (superior colliculus) NEP

Sequestered superior colliculus neurons in the neuroepithelium?

Skull and skin

Meninges (mostly *superarachnoid reticulum*)

Brain surface (pia, heavier line)

MESENCEPHALIC SUPERVENTRICLE (FUTURE AQUEDUCT)

MESENCEPHALIC TECTUM

Mesencephalic tectal (inferior colliculus) NEP

Nerve IV (trochlear, decussation)

INFERIOR COLLICULUS

ISTHMUS?

Isthmal NEP?

ISTHMAL CANAL?

Lateral lemniscus?

CEREBELLUM (HEMISPHERE)

CTF1 (fibers)

CTF2 (cells)

CTF3 (cells and fibers)

CTF4 (cells)

CTF5 (cells and fibers)

CTF6 (cells)

Sojourning and migrating deep neurons (CTF2+4)

Sojourning and migrating Purkinje cells (CTF6)

Cerebellar NEP

RHOMBENCEPHALIC SUPERVENTRICLE (FUTURE FOURTH VENTRICLE)

Dorsal rhombic lip

Medullary velum

Precerebellar NEP

Medullary NEP

Ventral rhombic lip

LATERAL RECESS

MEDULLA

Prepositus nucleus?

Medial longitudinal fasciculus?

Solitary nucleus and tract

Reticular formation

Raphe nuclear complex

Spinal nucleus (V)?

Inferior cerebellar peduncle?

Inferior olive complex

Expanding fetal rhombencephalic choroid plexus

Posterior intramural migratory stream (inferior olive neurons)

Nerve IX (glossopharyngeal)?

Medial lemniscus

Midline raphe glial system (provides structural support for brainstem flexures)

Glossopharyngeal ganglion (IX)?

ABBREVIATIONS:
NEP - Neuroepithelium
CTF - Cerebellar transitional field

FONT KEY:
VENTRICULAR DIVISIONS - CAPITALS
Germinal zone - Helvetica bold
Transient structure - Times bold italic
Permanent structure - Times Roman or **Bold**

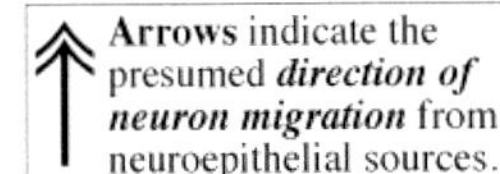
Arrows indicate the presumed ***direction of neuron migration*** from neuroepithelial sources.

PLATE 35A

CR 23 mm, GW 8.4, C966
Frontal/horizontal
Level 20: Section 171

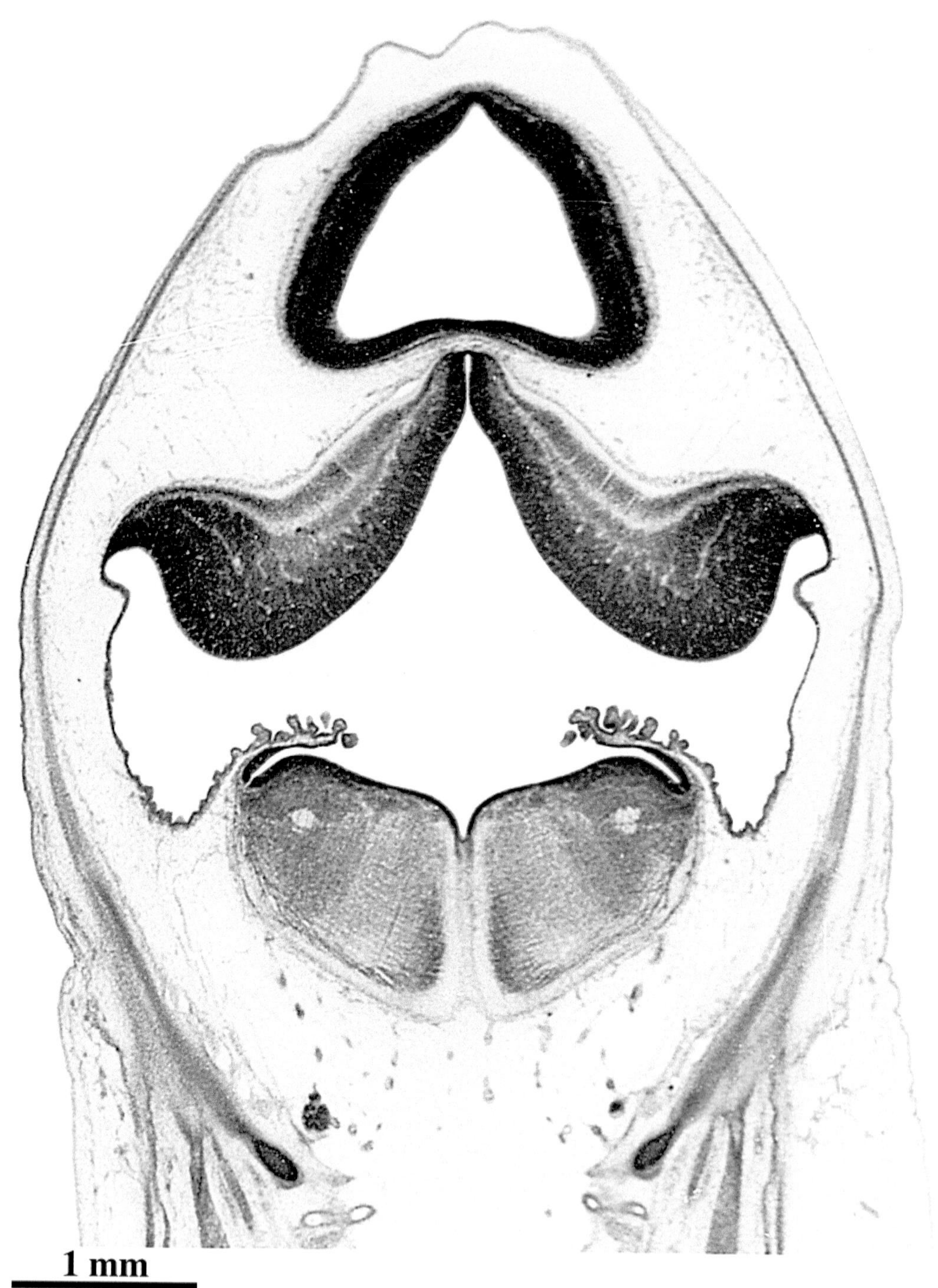
1 mm

PLATE 35B

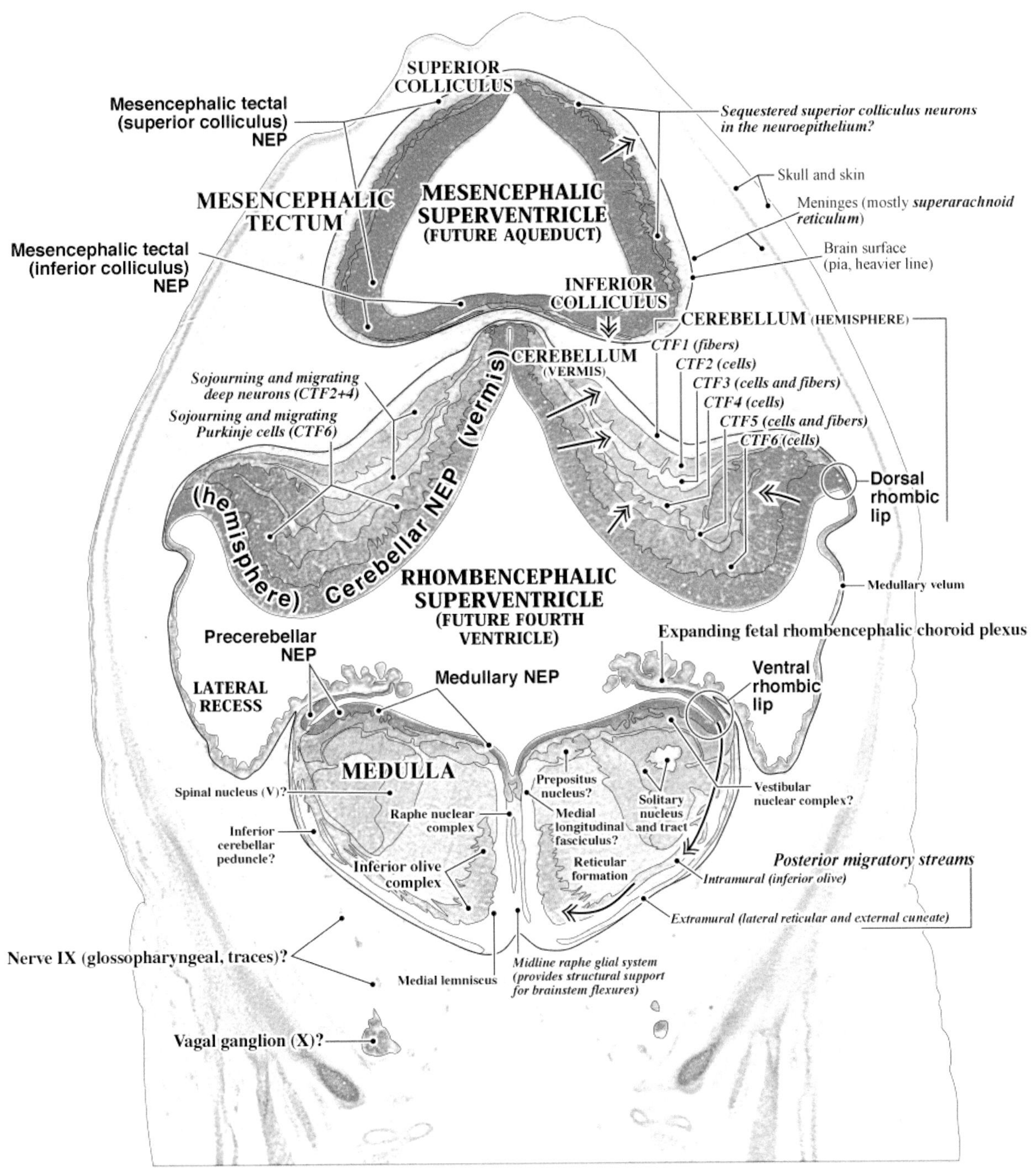
SUPERIOR COLLICULUS
Mesencephalic tectal (superior colliculus) NEP
Sequestered superior colliculus neurons in the neuroepithelium?
Skull and skin
Meninges (mostly superarachnoid reticulum)
Brain surface (pia, heavier line)
MESENCEPHALIC TECTUM
MESENCEPHALIC SUPERVENTRICLE (FUTURE AQUEDUCT)
Mesencephalic tectal (inferior colliculus) NEP
INFERIOR COLLICULUS
CEREBELLUM (HEMISPHERE)
CEREBELLUM (VERMIS)
CTF1 (fibers)
CTF2 (cells)
CTF3 (cells and fibers)
CTF4 (cells)
CTF5 (cells and fibers)
CTF6 (cells)
Sojourning and migrating deep neurons (CTF2+4)
Sojourning and migrating Purkinje cells (CTF6)
(vermis)
(hemisphere) Cerebellar NEP
Dorsal rhombic lip
Medullary velum
RHOMBENCEPHALIC SUPERVENTRICLE (FUTURE FOURTH VENTRICLE)
Expanding fetal rhombencephalic choroid plexus
Precerebellar NEP
Medullary NEP
Ventral rhombic lip
LATERAL RECESS
MEDULLA
Spinal nucleus (V)?
Raphe nuclear complex
Prepositus nucleus?
Medial longitudinal fasciculus?
Solitary nucleus and tract
Vestibular nuclear complex?
Inferior cerebellar peduncle?
Inferior olive complex
Reticular formation
Posterior migratory streams
Intramural (inferior olive)
Extramural (lateral reticular and external cuneate)
Nerve IX (glossopharyngeal, traces)?
Medial lemniscus
Midline raphe glial system (provides structural support for brainstem flexures)
Vagal ganglion (X)?

ABBREVIATIONS:
NEP - Neuroepithelium
CTF - Cerebellar transitional field

FONT KEY:
VENTRICULAR DIVISIONS - CAPITALS
Germinal zone - Helvetica bold
Transient structure - Times bold italic
Permanent structure - Times Roman or **Bold**

Arrows indicate the presumed ***direction of neuron migration*** from neuroepithelial sources.

PLATE 36A

CR 23 mm, GW 8.4, C966
Frontal/horizontal
Level 21: Section 179

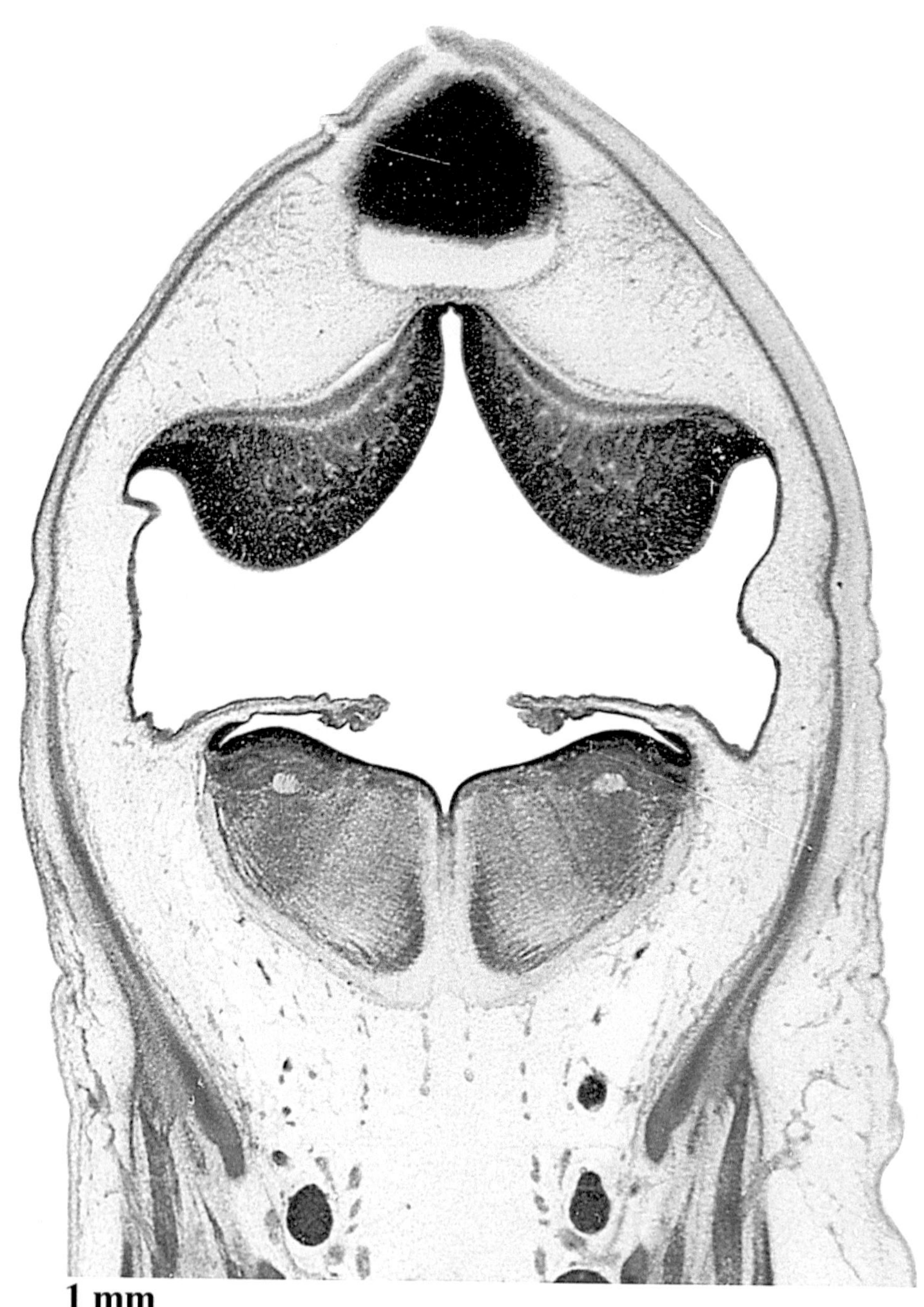
1 mm

PLATE 36B

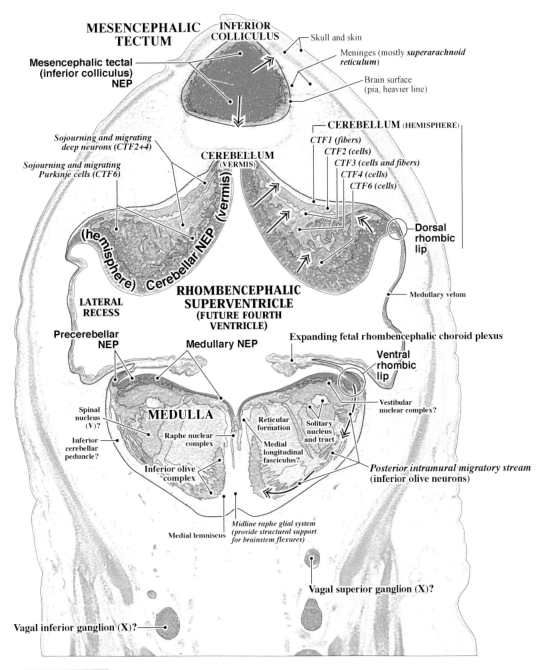

MESENCEPHALIC TECTUM
INFERIOR COLLICULUS
Skull and skin
Meninges (mostly ***superarachnoid reticulum***)
Mesencephalic tectal (inferior colliculus) NEP
Brain surface (pia, heavier line)
CEREBELLUM (HEMISPHERE)
Sojourning and migrating deep neurons (CTF2+4)
CTF1 (fibers)
CTF2 (cells)
CEREBELLUM (VERMIS)
Sojourning and migrating Purkinje cells (CTF6)
CTF3 (cells and fibers)
CTF4 (cells)
CTF6 (cells)
(vermis)
Dorsal rhombic lip
(hemisphere) Cerebellar NEP
RHOMBENCEPHALIC SUPERVENTRICLE (FUTURE FOURTH VENTRICLE)
Medullary velum
LATERAL RECESS
Precerebellar NEP
Medullary NEP
Expanding fetal rhombencephalic choroid plexus
Ventral rhombic lip
Vestibular nuclear complex?
Spinal nucleus (V)?
MEDULLA
Reticular formation
Solitary nucleus and tract
Inferior cerebellar peduncle?
Raphe nuclear complex
Medial longitudinal fasciculus?
Inferior olive complex
Posterior intramural migratory stream **(inferior olive neurons)**
Medial lemniscus
Midline raphe glial system (provide structural support for brainstem flexures)
Vagal superior ganglion (X)?
Vagal inferior ganglion (X)?

ABBREVIATIONS:
NEP - Neuroepithelium
CTF - Cerebellar transitional field

FONT KEY:
VENTRICULAR DIVISIONS - CAPITALS
Germinal zone - Helvetica bold
Transient structure - Times bold italic
Permanent structure - Times Roman or **Bold**

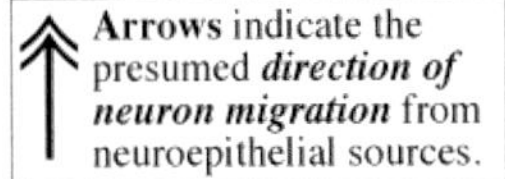
Arrows indicate the presumed ***direction of neuron migration*** from neuroepithelial sources.

FIGURE 22 GW8.4, CR23 mm, C966, COMPUTER-AIDED 3-D RECONSTRUCTION

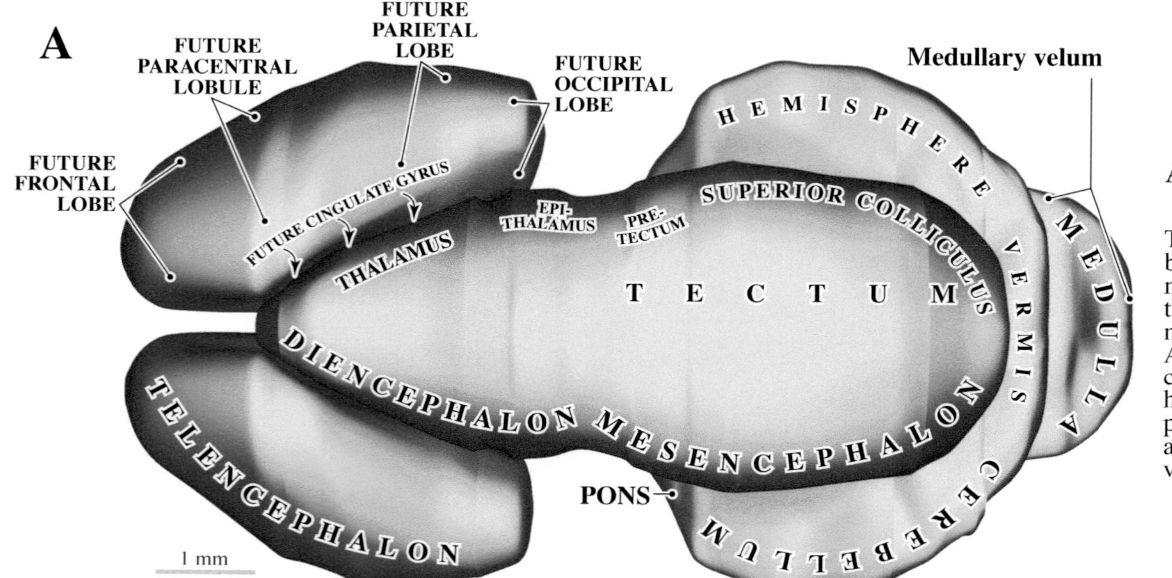

A
FUTURE FRONTAL LOBE
FUTURE PARACENTRAL LOBULE
FUTURE PARIETAL LOBE
FUTURE OCCIPITAL LOBE
Medullary velum
FUTURE CINGULATE GYRUS
THALAMUS
EPI-THALAMUS
PRE-TECTUM
SUPERIOR COLLICULUS
TECTUM
HEMISPHERE
VERMIS
MEDULLA
CEREBELLUM
DIENCEPHALON
MESENCEPHALON
TELENCEPHALON
PONS
1 mm

A. TOP VIEW OF THE BRAIN SURFACE

The observer is looking straight down on the top of the brain. Note the small size of the telencephalon. In a mature brain, the only structure visible from the top is the telencephalon. At GW 8.4, the telencephalon does not yet cover the dorsal surface of the diencephalon. All of the mesencephalic tectum is visible. The cerebellum appears as a wide ledge forming a horseshoe-shaped understory beneath the tectum. The pons is only partly visible connecting to the anteroventral edge of the cerebellum. The medullary velum is all that is visible of the medulla.

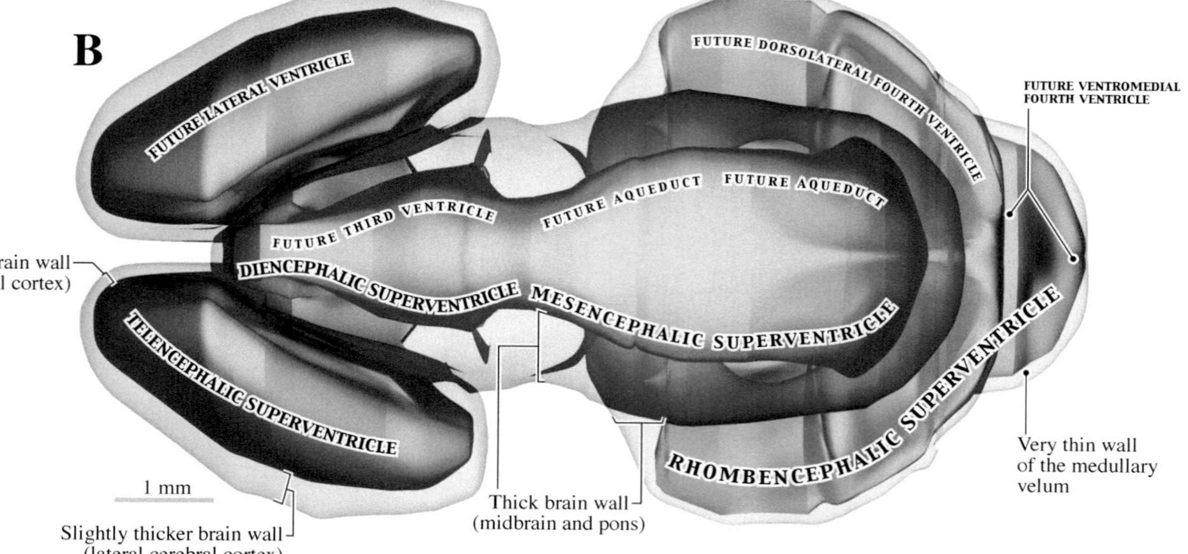

B
FUTURE LATERAL VENTRICLE
FUTURE DORSOLATERAL FOURTH VENTRICLE
FUTURE VENTROMEDIAL FOURTH VENTRICLE
FUTURE THIRD VENTRICLE
FUTURE AQUEDUCT
FUTURE AQUEDUCT
DIENCEPHALIC SUPERVENTRICLE
MESENCEPHALIC SUPERVENTRICLE
TELENCEPHALIC SUPERVENTRICLE
RHOMBENCEPHALIC SUPERVENTRICLE
Thin brain wall (medial cerebral cortex)
Slightly thicker brain wall (lateral cerebral cortex)
Thick brain wall (midbrain and pons)
Very thin wall of the medullary velum
1 mm

B. TOP VIEW SHOWING THE SUPERVENTRICLES

A substantial portion of the brain's volume is occupied by the ventricles. In a mature brain, the ventricles are small, central cavities. At GW 8.4, the diencephalic and mesencephalic superventricles are already narrowing to resemble their adult shapes, and the brain wall is thickest in that region. Most of the telencephalon is filled with the paired telencephalic superventricles, but note that the brain wall is thicker laterally than medially in accordance with the ventrolateral (earlier maturing) to dorsomedial (later maturing) developmental gradient. The dark areas within the brain wall are caused by looking through more than one thickness.

FIGURE 23 GW8.4, CR23 mm, C966, COMPUTER-AIDED 3-D RECONSTRUCTION

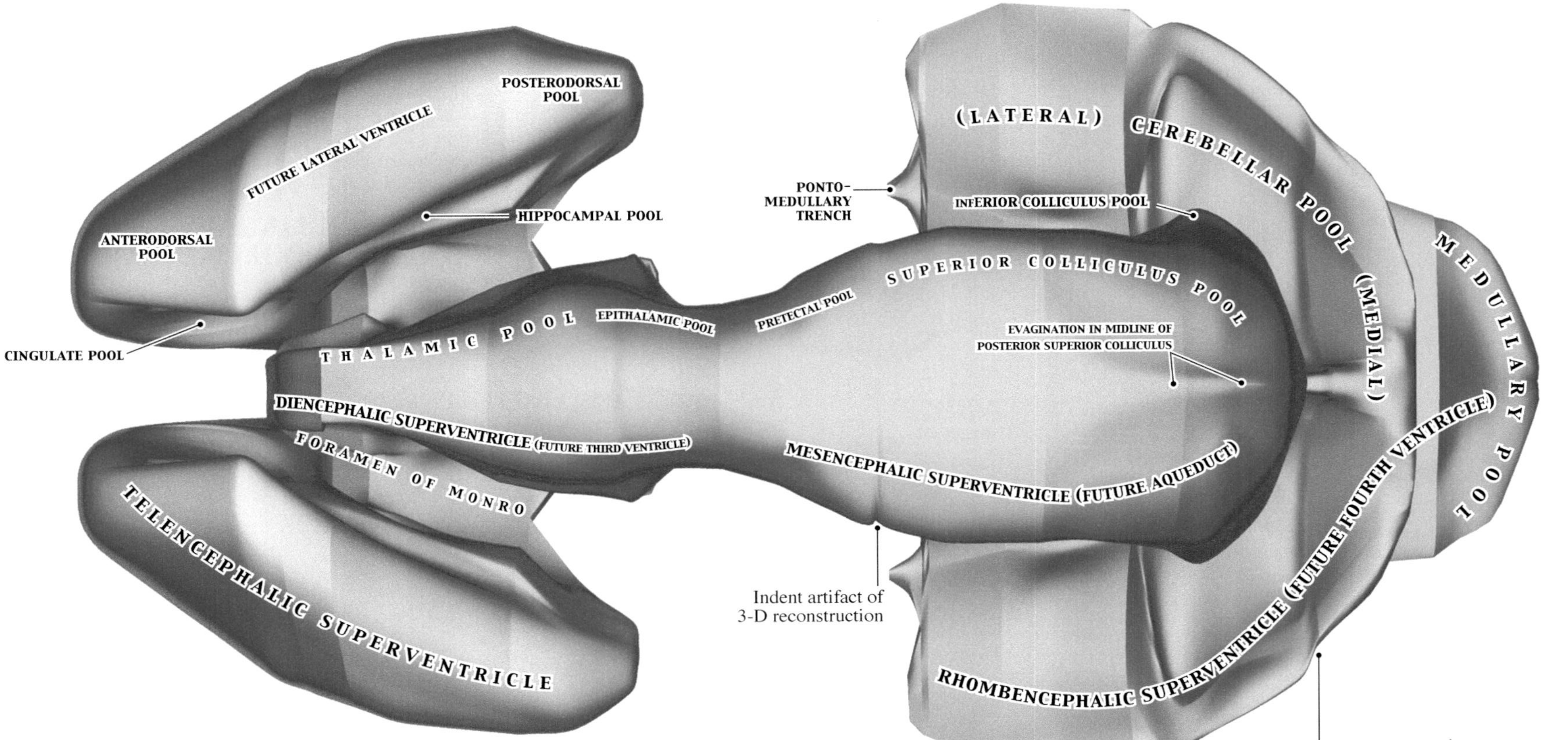
POSTERODORSAL POOL
FUTURE LATERAL VENTRICLE
HIPPOCAMPAL POOL
ANTERODORSAL POOL
CINGULATE POOL
THALAMIC POOL
EPITHALAMIC POOL
DIENCEPHALIC SUPERVENTRICLE (FUTURE THIRD VENTRICLE)
FORAMEN OF MONRO
TELENCEPHALIC SUPERVENTRICLE
PONTO-MEDULLARY TRENCH
PRETECTAL POOL
(LATERAL)
CEREBELLAR POOL
(MEDIAL)
INFERIOR COLLICULUS POOL
SUPERIOR COLLICULUS POOL
EVAGINATION IN MIDLINE OF POSTERIOR SUPERIOR COLLICULUS
MEDULLARY POOL
MESENCEPHALIC SUPERVENTRICLE (FUTURE AQUEDUCT)
RHOMBENCEPHALIC SUPERVENTRICLE (FUTURE FOURTH VENTRICLE)
Indent artifact of 3-D reconstruction
1 mm
Indent artifacts of 3-D reconstruction

TOP VIEW OF THE EXPOSED SUPERVENTRICLES

Enlarged view of **Figure 22B** showing only the ventricles. Because the ventricles contain fluid, subdivisions are called "pools." The various pools are named according to adjacent structures, which often produce evaginations into the ventricle (such as the cingulate and hippocampal pools). The complex shape of the brain ventricles is primarily a result of (1) localized differential proliferation in the adjacent neuroepithelium and (2) flexures in the diencephalon, midbrain, pons, and medulla during development.

FIGURE 24

GW8.4, CR23 mm, C966, COMPUTER-AIDED 3-D RECONSTRUCTION

A
FUTURE INSULAR AREA AND PRIMARY OLFACTORY CORTEX
FUTURE TEMPORAL LOBE
FUTURE ORBITO-FRONTAL AREA
FUTURE PREFRONTAL AREA
BASAL TELEN-CEPHALON
PREOPTIC AREA
SEPTUM
OPTIC RECESS
OLFACTORY BULB
TELENCEPHALON
HYPO-THALAMUS
INFUNDIBULUM
DIENCEPHALON
PONS
HEMISPHERE
CEREBELLUM
UPPER MEDULLA
Artifact of 3-D reconstruction
LOWER MEDULLA
SPINAL CORD
Artifact of 3-D reconstruction
1 mm

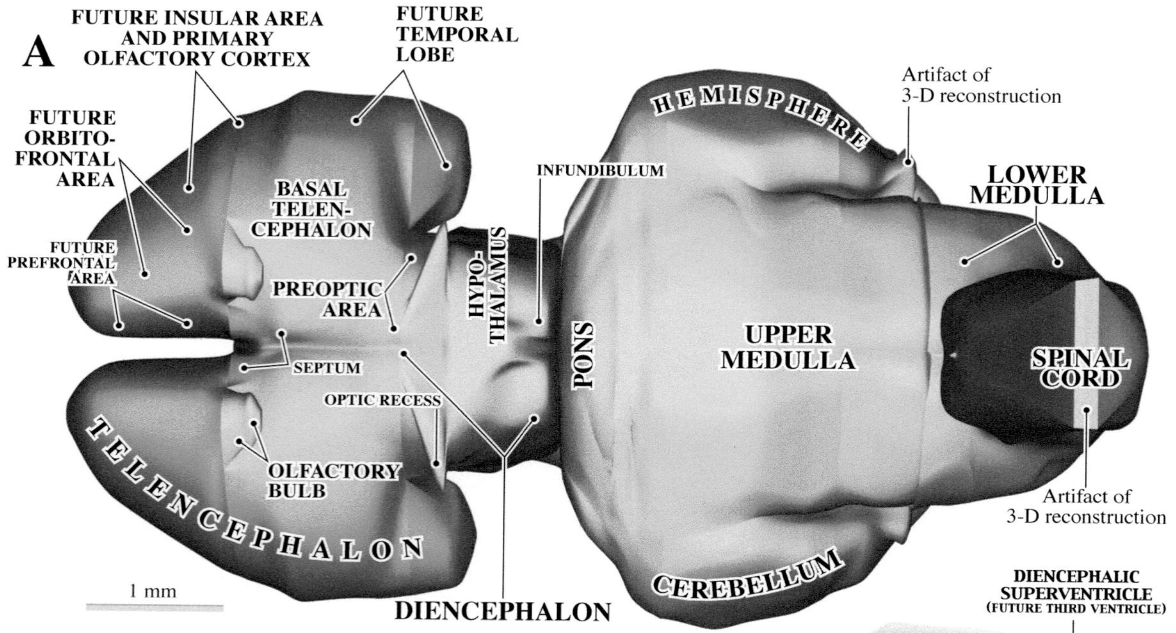

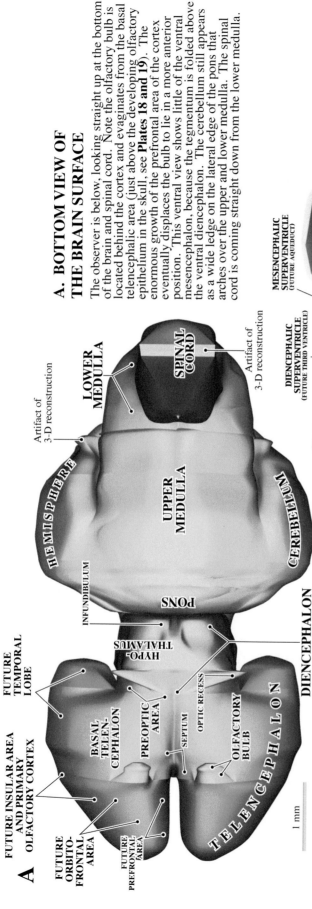

A. BOTTOM VIEW OF THE BRAIN SURFACE

The observer is below, looking straight up at the bottom of the brain and spinal cord. Note the olfactory bulb is located behind the cortex and evaginates from the basal telencephalic area (just above the developing olfactory epithelium in the skull, see **Plates 18 and 19**). The enormous growth of the prefrontal area of the cortex eventually displaces the bulb to lie in a more anterior position. This ventral view shows little of the ventral mesencephalon, because the tegmentum is folded above the ventral diencephalon. The cerebellum still appears as a wide ledge on the lateral edge of the pons that arches over the upper and lower medulla. The spinal cord is coming straight down from the lower medulla.

B
Thin brain wall (medial cerebral cortex)
FUTURE LATERAL VENTRICLE
TELENCEPHALIC SUPERVENTRICLE
1 mm
Slightly thicker brain wall (ventrolateral cerebral cortex)
DIENCEPHALIC SUPERVENTRICLE (FUTURE THIRD VENTRICLE)
MESENCEPHALIC SUPERVENTRICLE (FUTURE AQUEDUCT)
Thick brain wall (pons)
FUTURE LATERAL RECESS OF THE FOURTH VENTRICLE
FUTURE ANTEROVENTRAL FOURTH VENTRICLE
RHOMBENCEPHALIC SUPERVENTRICLE
FUTURE POSTEROVENTRAL FOURTH VENTRICLE
CENTRAL CANAL (SPINAL CORD)

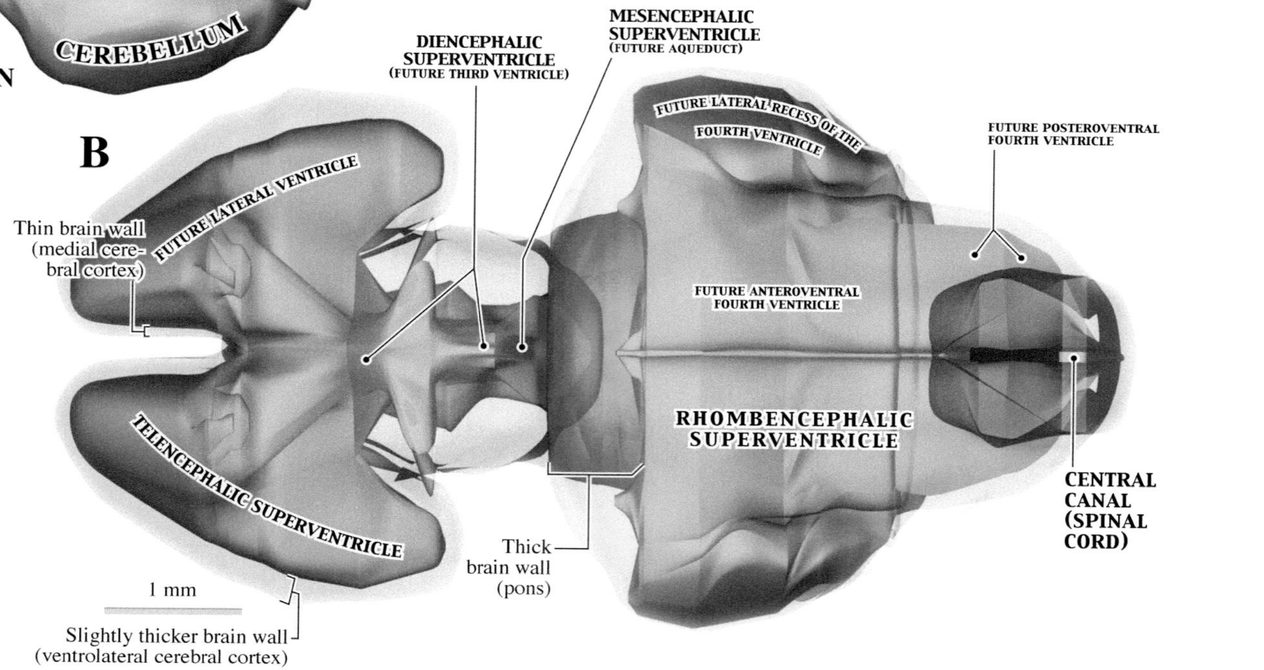

B. BOTTOM VIEW SHOWING THE SUPERVENTRICLES

The brain wall is thickest in the pons from this viewpoint because the pontine parenchyma is in front of the anterior edge of the rhombencephalic superventricle. In the telencephalon, the lateral brain wall is thicker than the medial wall in accordance with the ventrolateral (earlier maturing) to dorsomedial (later maturing) developmental gradient. As in **Figure 22B**, the dark areas within the brain wall are caused by looking through more than one thickness.

FIGURE 25

GW8.4, CR23 mm, C966, COMPUTER-AIDED 3-D RECONSTRUCTION

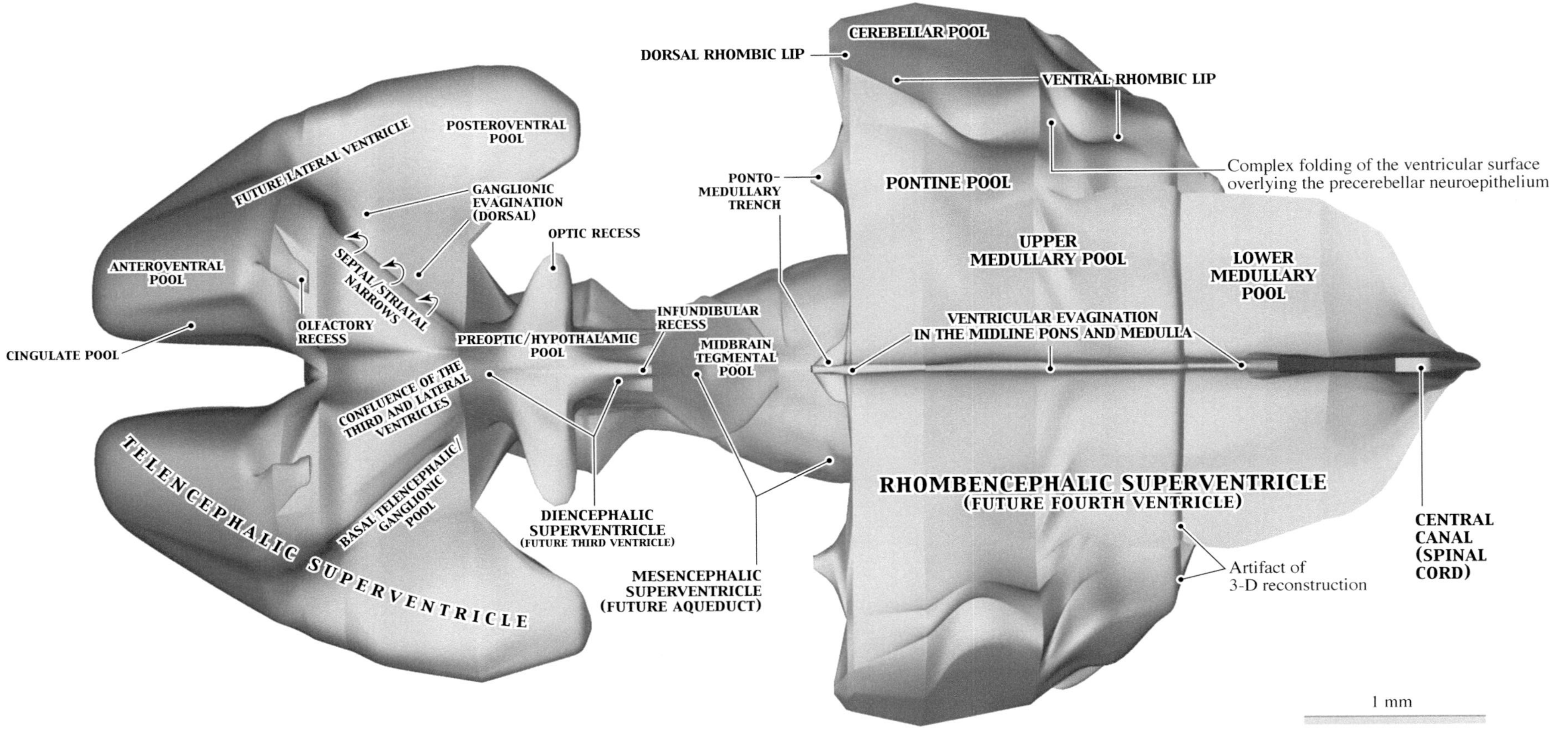

CEREBELLAR POOL
DORSAL RHOMBIC LIP
VENTRAL RHOMBIC LIP
POSTEROVENTRAL POOL
FUTURE LATERAL VENTRICLE
GANGLIONIC EVAGINATION (DORSAL)
PONTO-MEDULLARY TRENCH
PONTINE POOL
Complex folding of the ventricular surface overlying the precerebellar neuroepithelium
OPTIC RECESS
SEPTAL/STRIATAL NARROWS
ANTEROVENTRAL POOL
UPPER MEDULLARY POOL
LOWER MEDULLARY POOL
OLFACTORY RECESS
INFUNDIBULAR RECESS
VENTRICULAR EVAGINATION IN THE MIDLINE PONS AND MEDULLA
CINGULATE POOL
PREOPTIC/HYPOTHALAMIC POOL
MIDBRAIN TEGMENTAL POOL
CONFLUENCE OF THE THIRD AND LATERAL VENTRICLES
TELENCEPHALIC SUPERVENTRICLE
BASAL TELENCEPHALIC/GANGLIONIC POOL
RHOMBENCEPHALIC SUPERVENTRICLE (FUTURE FOURTH VENTRICLE)
DIENCEPHALIC SUPERVENTRICLE (FUTURE THIRD VENTRICLE)
CENTRAL CANAL (SPINAL CORD)
Artifact of 3-D reconstruction
MESENCEPHALIC SUPERVENTRICLE (FUTURE AQUEDUCT)
1 mm

BOTTOM VIEW OF THE EXPOSED SUPERVENTRICLES

Enlarged view of **Figure 24B** showing only the ventricles. The complex evaginations of the ventricular surface indicate the heterogeneous nature of the neuroepithelium that lines different ventricular divisions.

FIGURE 26

GW8.4, CR23 mm, C966, COMPUTER-AIDED 3-D RECONSTRUCTION

A. SIDE VIEW OF THE BRAIN SURFACE

The observer is beside the model, looking at the lateral surface of the brain and upper cervical spinal cord. Note the posterior-facing olfactory bulb beneath the basal telencephaon. This lateral view shows most clearly all of the flexures in the brainstem.

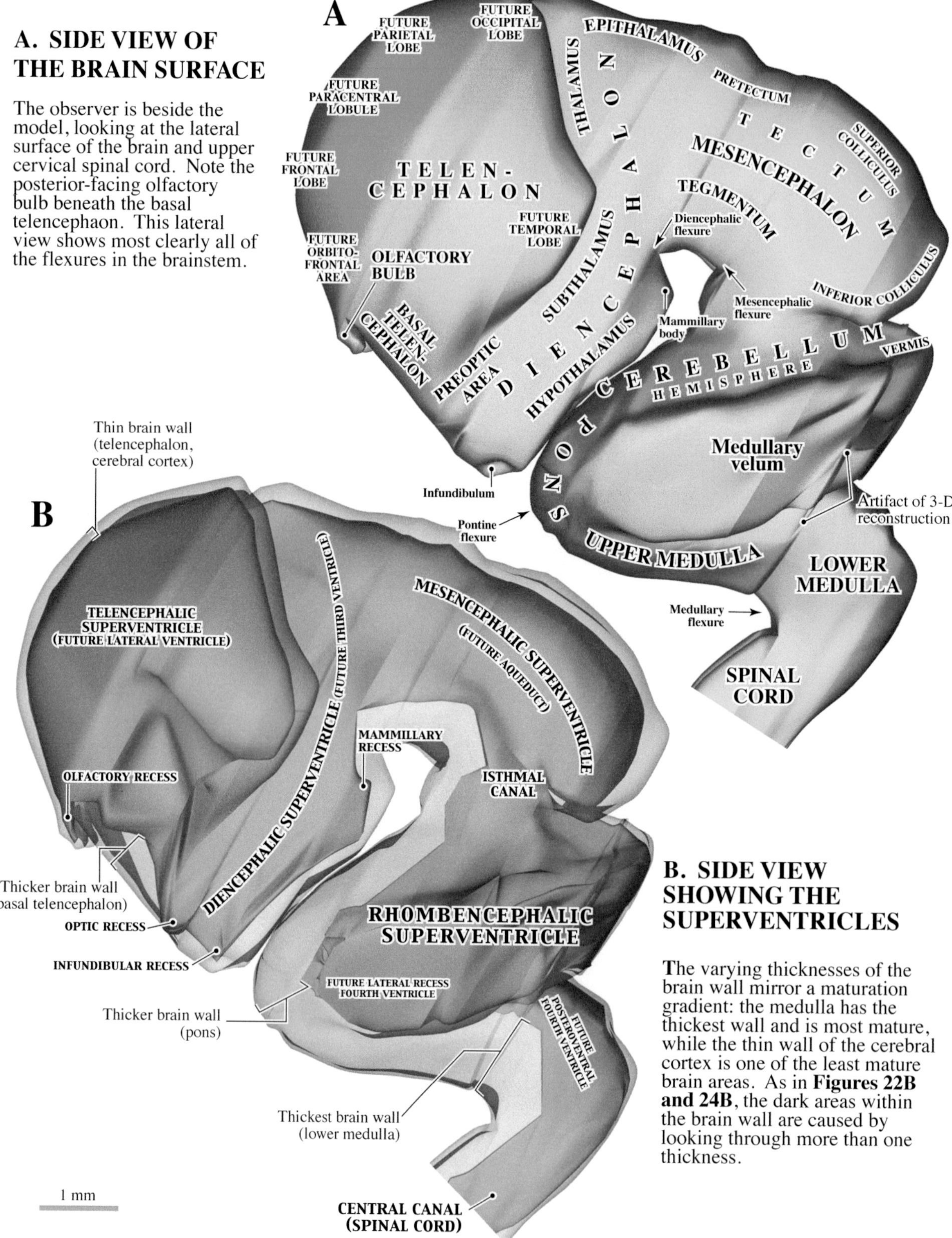
A
FUTURE PARIETAL LOBE
FUTURE OCCIPITAL LOBE
FUTURE PARACENTRAL LOBULE
FUTURE FRONTAL LOBE
TELEN-CEPHALON
FUTURE TEMPORAL LOBE
FUTURE ORBITO-FRONTAL AREA
OLFACTORY BULB
BASAL TELEN-CEPHALON
PREOPTIC AREA
THALAMUS
EPITHALAMUS
PRETECTUM
TECTUM
SUPERIOR COLLICULUS
MESENCEPHALON
TEGMENTUM
DIENCEPHALON
SUBTHALAMUS
HYPOTHALAMUS
Diencephalic flexure
Mammillary body
Mesencephalic flexure
INFERIOR COLLICULUS
CEREBELLUM
VERMIS
HEMISPHERE
PONS
Medullary velum
Infundibulum
Pontine flexure
Artifact of 3-D reconstruction
UPPER MEDULLA
LOWER MEDULLA
Medullary flexure
SPINAL CORD
B
Thin brain wall (telencephalon, cerebral cortex)
TELENCEPHALIC SUPERVENTRICLE (FUTURE LATERAL VENTRICLE)
DIENCEPHALIC SUPERVENTRICLE (FUTURE THIRD VENTRICLE)
MESENCEPHALIC SUPERVENTRICLE
(FUTURE AQUEDUCT)
MAMMILLARY RECESS
ISTHMAL CANAL
OLFACTORY RECESS
Thicker brain wall (basal telencephalon)
OPTIC RECESS
INFUNDIBULAR RECESS
RHOMBENCEPHALIC SUPERVENTRICLE
FUTURE LATERAL RECESS FOURTH VENTRICLE
FUTURE POSTEROVENTRAL FOURTH VENTRICLE
Thicker brain wall (pons)
Thickest brain wall (lower medulla)
CENTRAL CANAL (SPINAL CORD)
1 mm

B. SIDE VIEW SHOWING THE SUPERVENTRICLES

The varying thicknesses of the brain wall mirror a maturation gradient: the medulla has the thickest wall and is most mature, while the thin wall of the cerebral cortex is one of the least mature brain areas. As in **Figures 22B and 24B**, the dark areas within the brain wall are caused by looking through more than one thickness.

GW8.4, CR23 mm, C966, COMPUTER-AIDED 3-D RECONSTRUCTION

FIGURE 27

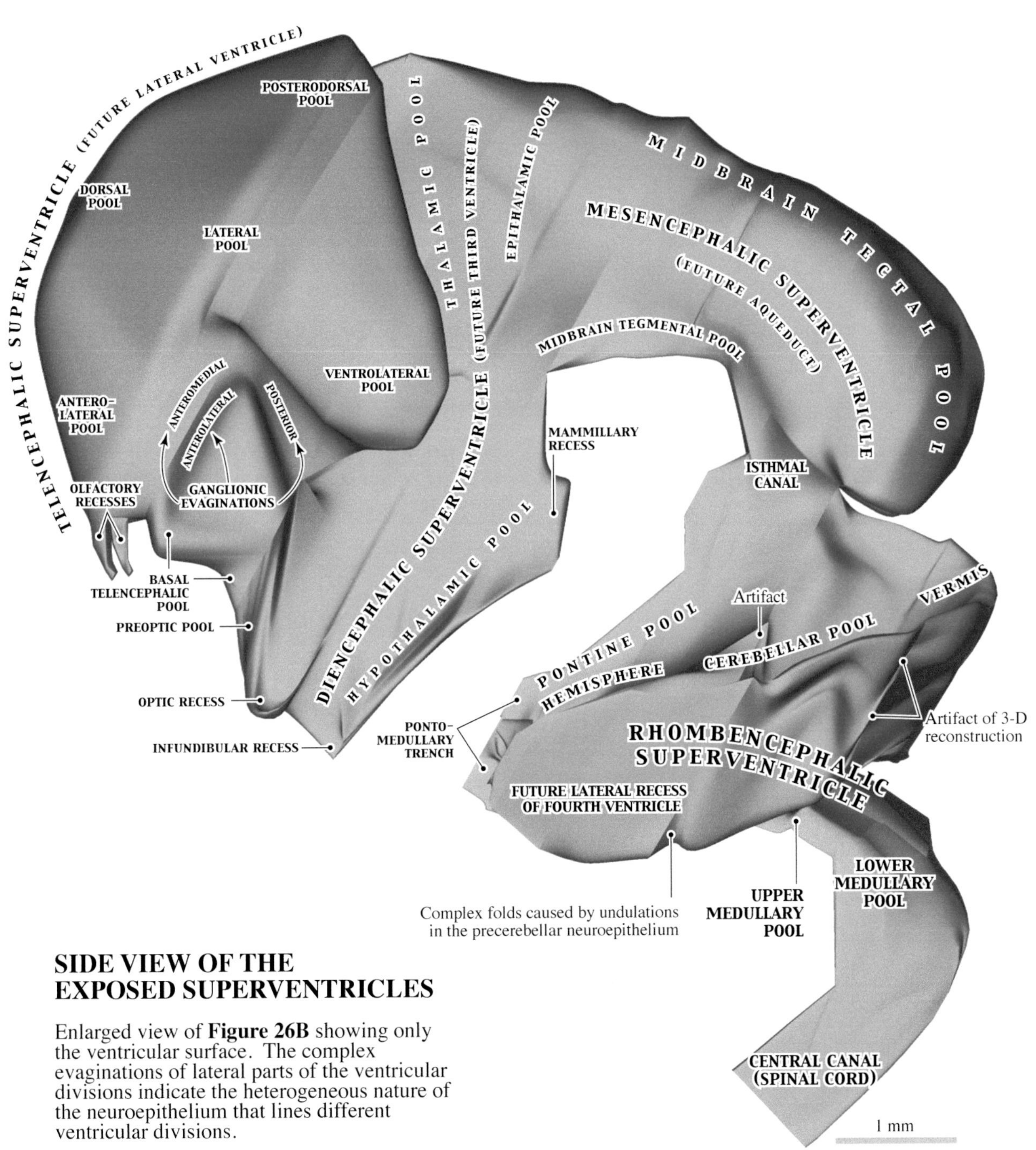
TELENCEPHALIC SUPERVENTRICLE (FUTURE LATERAL VENTRICLE)
POSTERODORSAL POOL
DORSAL POOL
LATERAL POOL
ANTERO-LATERAL POOL
VENTROLATERAL POOL
ANTEROMEDIAL
ANTEROLATERAL
POSTERIOR
GANGLIONIC EVAGINATIONS
OLFACTORY RECESSES
BASAL TELENCEPHALIC POOL
PREOPTIC POOL
OPTIC RECESS
INFUNDIBULAR RECESS
THALAMIC POOL
DIENCEPHALIC SUPERVENTRICLE (FUTURE THIRD VENTRICLE)
EPITHALAMIC POOL
HYPOTHALAMIC POOL
MAMMILLARY RECESS
MIDBRAIN TECTAL POOL
MESENCEPHALIC SUPERVENTRICLE (FUTURE AQUEDUCT)
MIDBRAIN TEGMENTAL POOL
ISTHMAL CANAL
VERMIS
Artifact
PONTINE POOL
CEREBELLAR POOL
HEMISPHERE
PONTO-MEDULLARY TRENCH
RHOMBENCEPHALIC SUPERVENTRICLE
Artifact of 3-D reconstruction
FUTURE LATERAL RECESS OF FOURTH VENTRICLE
Complex folds caused by undulations in the precerebellar neuroepithelium
UPPER MEDULLARY POOL
LOWER MEDULLARY POOL
CENTRAL CANAL (SPINAL CORD)
1 mm

SIDE VIEW OF THE EXPOSED SUPERVENTRICLES

Enlarged view of **Figure 26B** showing only the ventricular surface. The complex evaginations of lateral parts of the ventricular divisions indicate the heterogeneous nature of the neuroepithelium that lines different ventricular divisions.

FIGURE 28

GW8.4, CR23 mm, C966, COMPUTER-AIDED 3-D RECONSTRUCTION OF THE LATERAL TELENCEPHALIC NEUROEPITHELIUM

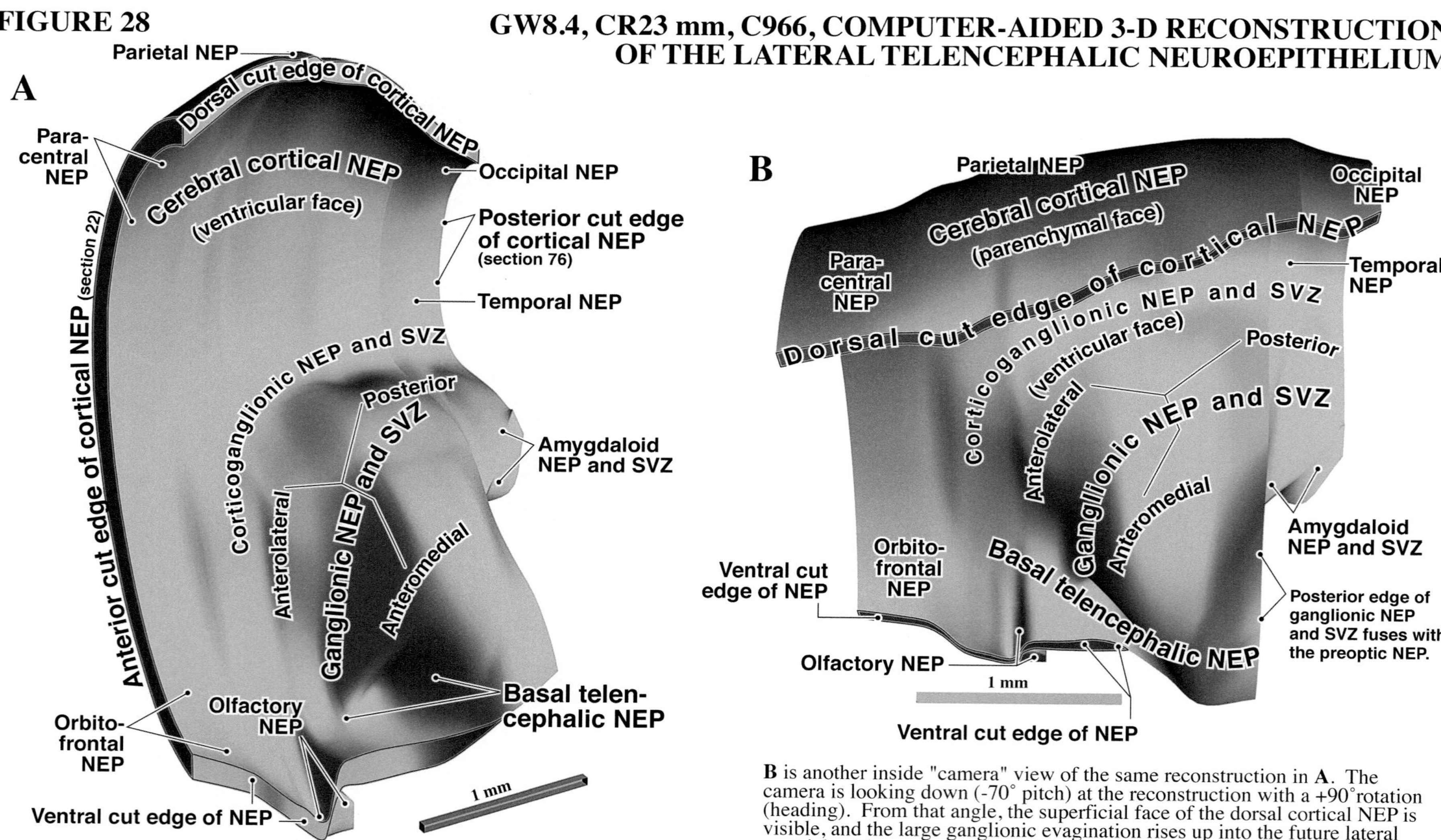
A
Parietal NEP
Dorsal cut edge of cortical NEP
Para-
central
NEP
Cerebral cortical NEP
(ventricular face)
Occipital NEP
Posterior cut edge
of cortical NEP
(section 76)
Temporal NEP
Anterior cut edge of cortical NEP (section 22)
Corticoganglionic NEP and SVZ
Posterior
Ganglionic NEP and SVZ
Anterolateral
Anteromedial
Amygdaloid
NEP and SVZ
Olfactory
NEP
Basal telen-
cephalic NEP
Orbito-
frontal
NEP
Ventral cut edge of NEP
1 mm
B
Parietal NEP
Cerebral cortical NEP
(parenchymal face)
Occipital
NEP
Dorsal cut edge of cortical NEP
Para-
central
NEP
Temporal
NEP
Corticoganglionic NEP and SVZ
(ventricular face)
Posterior
Anterolateral
Ganglionic NEP and SVZ
Anteromedial
Amygdaloid
NEP and SVZ
Orbito-
frontal
NEP
Basal telencephalic NEP
Ventral cut
edge of NEP
Posterior edge of
ganglionic NEP
and SVZ fuses with
the preoptic NEP.
Olfactory NEP
1 mm
Ventral cut edge of NEP

A is an inside "camera" view of the lateral half of the telencephalic neuroepithelium and subventricular zone; the medial half is not reconstructed so that the ganglionic and olfactory evaginations are visible. The reconstruction includes sections 22 (between **Levels 1 and 2, Plates 17 and 18**) through 76 (**Level 7, Plate 22**). The camera is looking straight (0° pitch) at the reconstruction with a +60°rotation (heading). The large ganglionic evagination is evident in the ventrolateral NEP and SVZ.

B is another inside "camera" view of the same reconstruction in **A**. The camera is looking down (-70° pitch) at the reconstruction with a +90°rotation (heading). From that angle, the superficial face of the dorsal cortical NEP is visible, and the large ganglionic evagination rises up into the future lateral ventricle. The posterior edge of the ganglionic NEP and SVZ fuses with the preoptic NEP posterior to the foramen of Monro. See that transition between Level 5 (**Plate 20**) and Level 6 (**Plate 21**).

ABBREVIATIONS:
NEP - Neuroepithelium
SVZ - Subventricular zone

Neuroepithelial compartments labeled in Helvetica Bold.

GW8.4, CR23 mm, C966, COMPUTER-AIDED 3-D RECONSTRUCTION OF THE DIENCEPHALIC NEUROEPITHELIUM

FIGURE 29

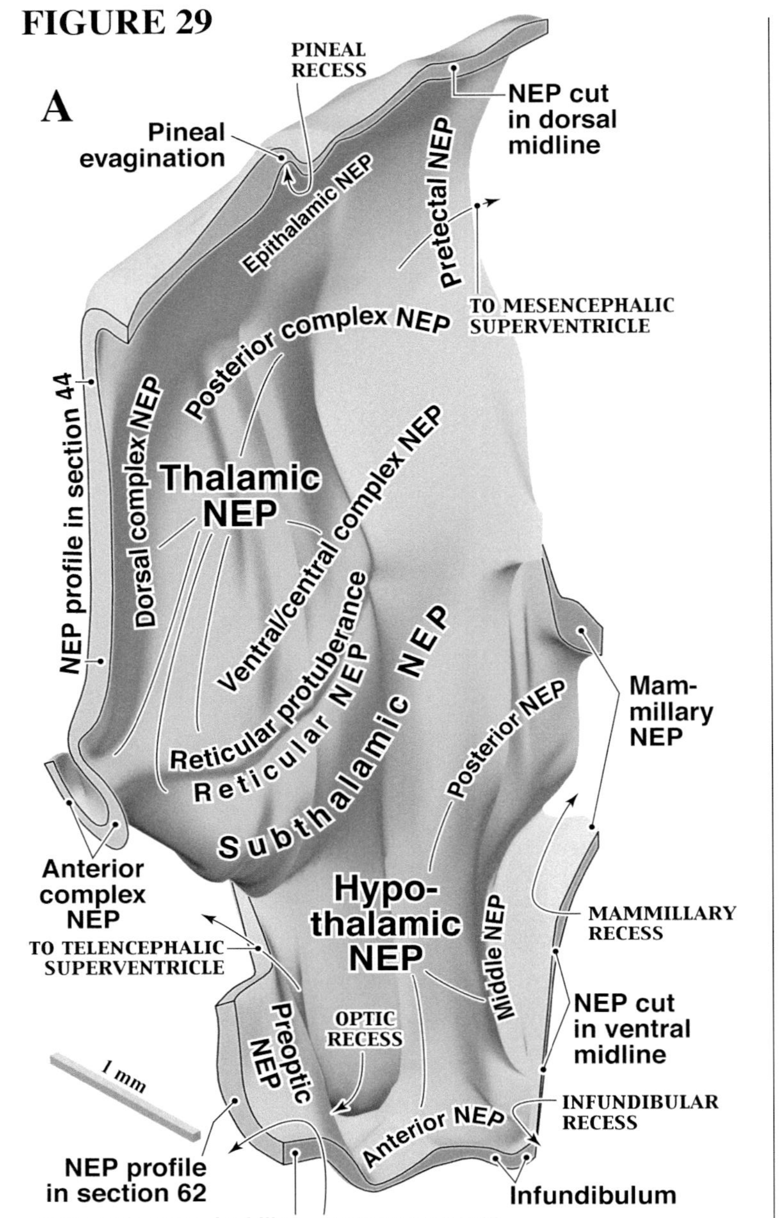
A
Pineal evagination
PINEAL RECESS
NEP cut in dorsal midline
Epithalamic NEP
Pretectal NEP
TO MESENCEPHALIC SUPERVENTRICLE
Posterior complex NEP
NEP profile in section 44
Dorsal complex NEP
Thalamic NEP
Ventral/central complex NEP
Reticular protuberance
Reticular NEP
Subthalamic NEP
Posterior NEP
Mam-millary NEP
Anterior complex NEP
TO TELENCEPHALIC SUPERVENTRICLE
Hypo-thalamic NEP
Middle NEP
MAMMILLARY RECESS
Preoptic NEP
OPTIC RECESS
NEP cut in ventral midline
1 mm
Anterior NEP
INFUNDIBULAR RECESS
NEP profile in section 62
Infundibulum
NEP cut in ventral midline
PREOPTIC RECESS
B
DIENCEPHALIC SUPERVENTRICLE (FUTURE THIRD VENTRICLE)
Pretectal/tectal NEP (profile in section 102)
Epithalamic NEP
Thalamic NEP
Posterior complex NEP
Tegmental NEP (profile in section 102)
Anterior complex NEP
Mammillary NEP (profile in section 112)
Optic recess NEP
Hypo-thalamic NEP
Middle NEP
1 mm

Two inside "camera" views of the right half of the diencephalic neuroepithelium that includes section 44 (near **Level 3, Plate 18**) through section 112 (**Level 12, Plate 19**); the NEP in all sections has been cut where it bridges the midline dorsally and ventrally so that its folds and undulations can be observed.

A. The camera views the front of the reconstruction with +45°heading and -25° pitch. From this angle, we see the ventricular face of the NEP.

B. The camera views the back of the reconstruction with 180° heading and -10° pitch. From this angle, we see some of the parenchymal face of the NEP. Note the prominent folding in the anterior complex NEP, the thick posterior complex NEP, and the back wall of the NEP lining the optic recess.

It is postulated that the multiple evaginations and invaginations of the diencephalic neuroepithelium are mosaic compartments that give rise to different thalamic, hypothalamic, and preoptic nuclei.

ABBREVIATION:
NEP - Neuroepithelium

Neuroepithelial compartments labeled in Helvetica Bold.
VENTRICULAR DIVISIONS LABELED IN CAPITALS.

FIGURE 30
GW8.4, CR23 mm, C966, COMPUTER-AIDED 3-D RECONSTRUCTION OF THE RHOMBENCEPHALIC NEUROEPITHELIUM

The reconstruction includes section 128 (**Level 15, Plate 30**) through section 193 (posterior to **Level 21, Plate 36**); the NEP in all sections has been cut where it bridges the midline dorsally and ventrally, and only the right half is shown. The camera is looking at the front of the NEP (0° heading, -10° pitch).

A shows both the roof and floor plates of the rhombencephalic NEP with orientation labels.

B shows the parenchymal face of the roof NEP bordering the differentiating zones of the pons, isthmus, and cerebellum). Since the edges between ***CTF6*** (Purkinje cell sojourn zone) and the **cerebellar NEP** are virtually indistinguishable, both are included in the reconstruction.

C shows the ventricular face of the floor NEP overlying the upper (adjacent to pons) and lower medulla (adjacent to spinal cord).

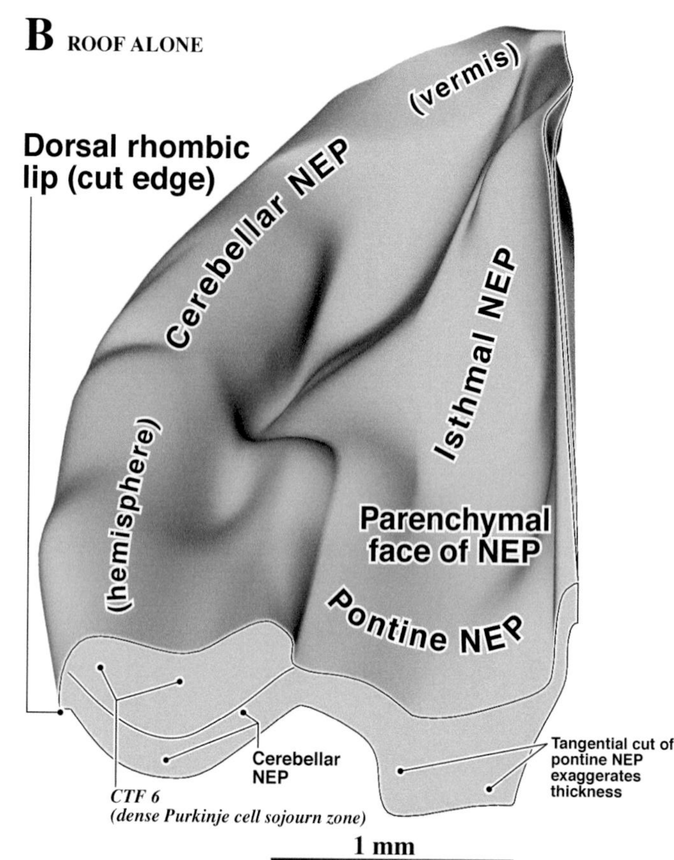
B ROOF ALONE
(vermis)
Dorsal rhombic lip (cut edge)
Cerebellar NEP
Isthmal NEP
(hemisphere)
Parenchymal face of NEP
Pontine NEP
Cerebellar NEP
Tangential cut of pontine NEP exaggerates thickness
CTF 6
(dense Purkinje cell sojourn zone)
1 mm

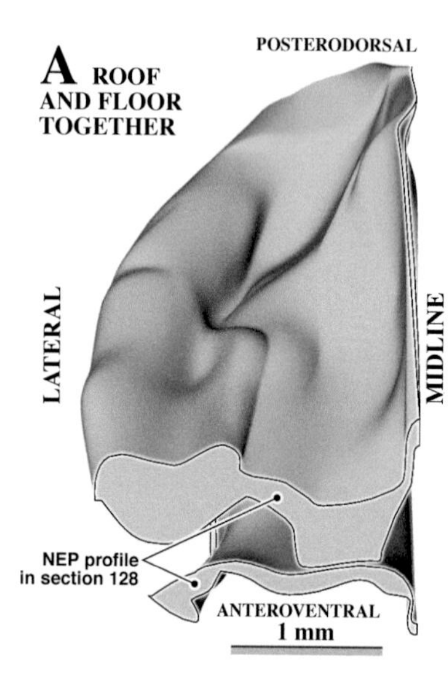
A ROOF AND FLOOR TOGETHER
POSTERODORSAL
LATERAL
MIDLINE
NEP profile in section 128
ANTEROVENTRAL
1 mm

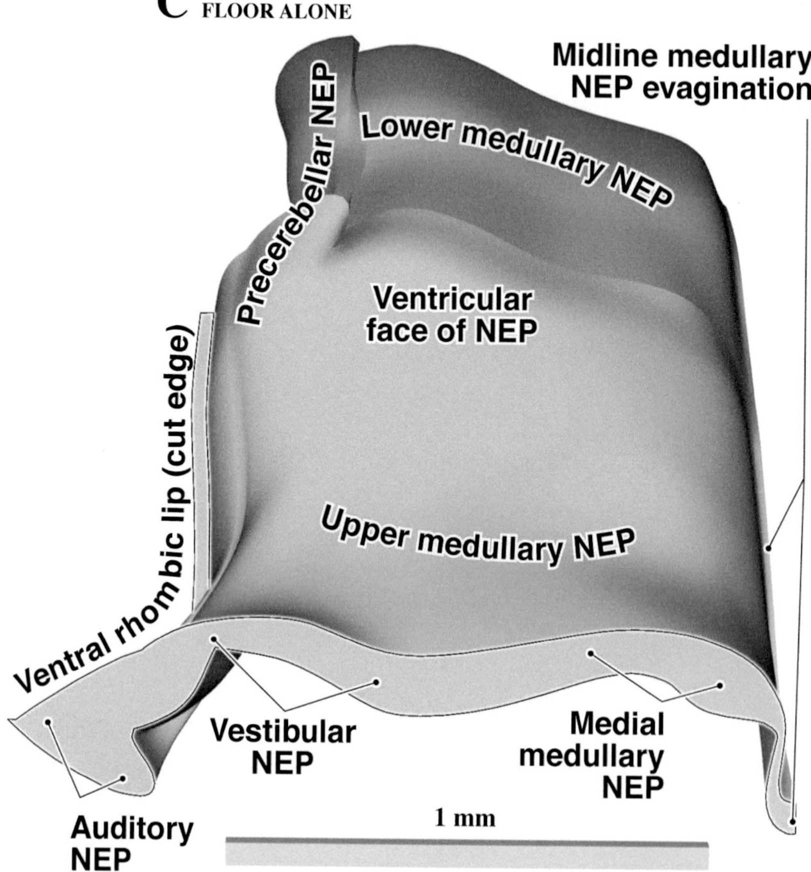
C FLOOR ALONE
Midline medullary NEP evagination
Precerebellar NEP
Lower medullary NEP
Ventricular face of NEP
Ventral rhombic lip (cut edge)
Upper medullary NEP
Vestibular NEP
Medial medullary NEP
Auditory NEP
1 mm

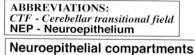

ABBREVIATIONS:
CTF - Cerebellar transitional field
NEP - Neuroepithelium

Neuroepithelial compartments labeled in Helvetica Bold.
Transient developmental structures labeled in Times Bold Italic

FIGURE 31
GW8.4, CR23 mm, C966, COMPUTER-AIDED 3-D RECONSTRUCTION OF THE RHOMBENCEPHALIC NEUROEPITHELIUM

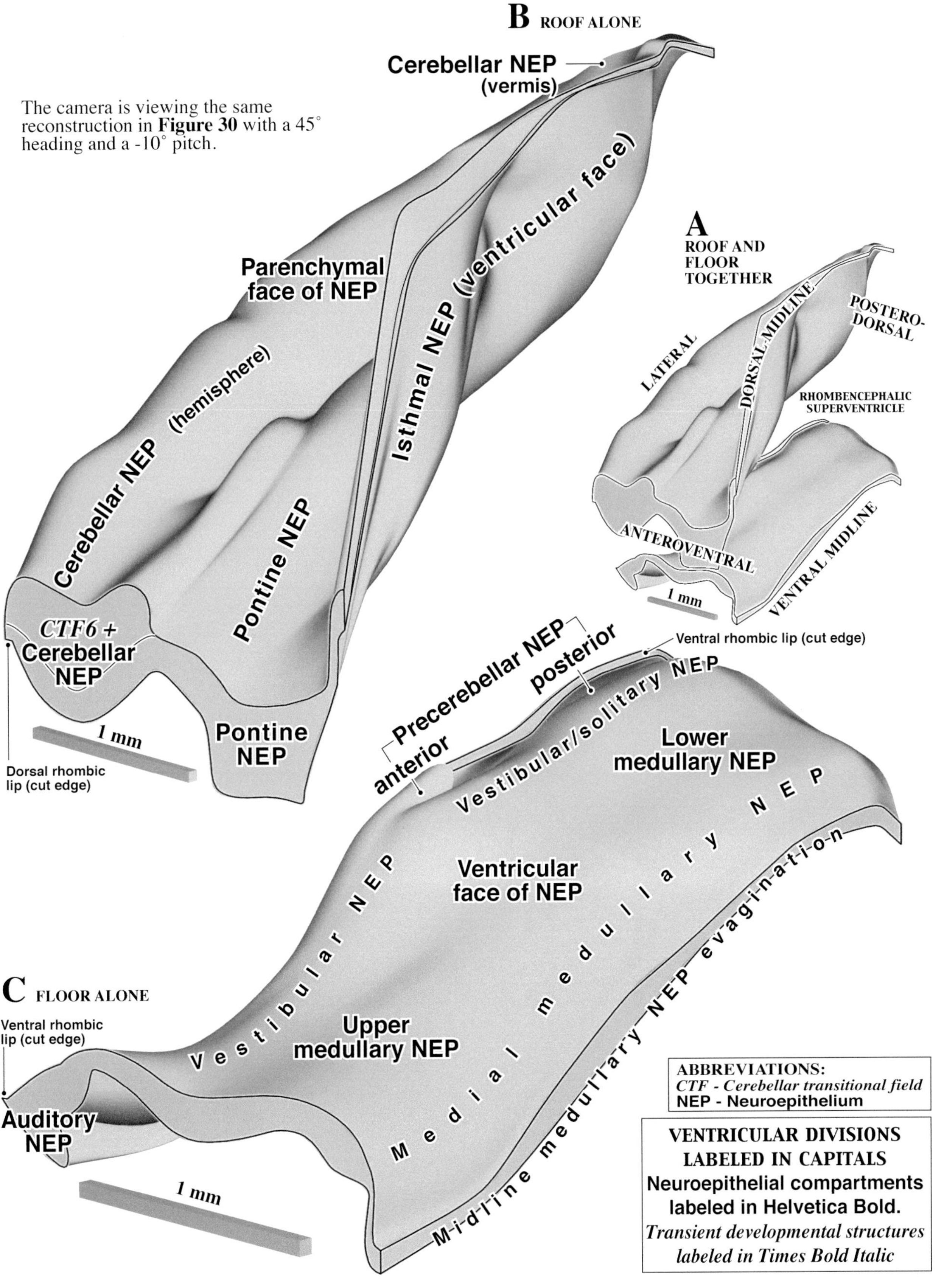
B ROOF ALONE
Cerebellar NEP
(vermis)
The camera is viewing the same reconstruction in **Figure 30** with a 45° heading and a -10° pitch.
Parenchymal face of NEP
Isthmal NEP (ventricular face)
Cerebellar NEP (hemisphere)
Pontine NEP
CTF6 + Cerebellar NEP
Pontine NEP
1 mm
Dorsal rhombic lip (cut edge)
A
ROOF AND FLOOR TOGETHER
POSTERO-DORSAL
DORSAL MIDLINE
LATERAL
RHOMBENCEPHALIC SUPERVENTRICLE
ANTEROVENTRAL
VENTRAL MIDLINE
1 mm
Ventral rhombic lip (cut edge)
Precerebellar NEP
posterior
anterior
Vestibular/solitary NEP
Lower medullary NEP
Medial medullary NEP
Ventricular face of NEP
Vestibular NEP
Midline medullary NEP evagination
Upper medullary NEP
C FLOOR ALONE
Ventral rhombic lip (cut edge)
Auditory NEP
1 mm
ABBREVIATIONS:
CTF - Cerebellar transitional field
NEP - Neuroepithelium
VENTRICULAR DIVISIONS LABELED IN CAPITALS
Neuroepithelial compartments labeled in Helvetica Bold.
Transient developmental structures labeled in Times Bold Italic